Lithium-ion Batteries - Thin Film for Energy Materials and Devices

Edited by Mitsunobu Sato, Li Lu and Hiroki Nagai

Published in London, United Kingdom

IntechOpen

Supporting open minds since 2005

Lithium-ion Batteries - Thin Film for Energy Materials and Devices
http://dx.doi.org/10.5772/intechopen.73346
Edited by Mitsunobu Sato, Li Lu and Hiroki Nagai

Contributors
Xiaogang Sun, Xu Li, Wei Chen, Jie Wang, Chengcheng Wei, Yapan Huang, Guodong Liang, Hao Hu, Rachid Yazami, Wafaa Nachit, Samira Touhtouh, Zahra Ramzi, Khalil Benkhouja, Yuanli Ding, Janina Molenda, Mitsunobu Sato, Hiroki Nagai, Philipus Nghilukunanye Hishimone, Dhamodaran Santhanagopalan, Silpasree S Jayasree, Shantikumar Nair

First published in London, United Kingdom, 2020 by IntechOpen
IntechOpen is the global imprint of INTECHOPEN LIMITED, registered in England and Wales, registration number: 11086078, 7th floor, 10 Lower Thames Street, London, EC3R 6AF, United Kingdom
Printed in Croatia

British Library Cataloguing-in-Publication Data
A catalogue record for this book is available from the British Library

Additional hard and PDF copies can be obtained from orders@intechopen.com

Lithium-ion Batteries - Thin Film for Energy Materials and Devices
Edited by Mitsunobu Sato, Li Lu and Hiroki Nagai
p. cm.
Print ISBN 978-1-78985-463-3
Online ISBN 978-1-78985-464-0
eBook (PDF) ISBN 978-1-83962-297-7

An electronic version of this book is freely available, thanks to the support of libraries working with Knowledge Unlatched. KU is a collaborative initiative designed to make high quality books Open Access for the public good. More information about the initiative and links to the Open Access version can be found at www.knowledgeunlatched.org

For EU product safety concerns:
IN TECH d.o.o., Prolaz Marije Krucifikse Kozulić 3, 51000 Rijeka, Croatia, info@intechopen.com or visit our website at intechopen.com.

Meet the editors

Prof. Dr. Sato has been a professor of Chemistry in the Faculty of Engineering, Kogakuin University of Technology and Engineering (KUTE) since 2002. Since 2015, he has been a professor of Applied Physics in the School of Advanced Engineering, and the President of KUTE. He has been engaged in research in the fields of materials science and nanotechnology.

Prof. Dr. Lu is a professor in the Department of Mechanical Engineering, National University of Singapore (NUS). Dr. Lu is heavily involved in the research of functional materials, mainly in two directions: i) materials for Li-ion rechargeable batteries, which include traditional bulk batteries and all-solid-state batteries, and supercapacitors, and ii) piezoelectric and ferroelectric materials.

Dr. Nagai has been an associate professor at the Department of Applied Physics, School of Advanced Engineering, Kogakuin University of Technology & Engineering (KUTE-Tokyo) since 2018. He has been engaged in research in the fields of materials science and nanotechnology.

Contents

Preface

The global society requires sustainable and improved technologies for producing, converting, and storing energy. Innovative materials to afford useful devices are the key that we urgently need in energy utilization. Among the competing batteries, the rechargeable lithium-ion battery can essentially provide the highest specific energy and energy densities. The number of devices that are operated with lithium-ion battery continuously increases, and our daily lives depend on their performance. Thin-film lithium-ion batteries have already been implemented in several devices, such as microchips, medical implants, and small-scale power suppliers. The demands of thin-film lithium-ion battery having better performances are increasing together with the growing number of applications. However, there are still many problems to be solved for the market.

This book provides recent research and trends for thin film materials relevant to energy utilization. The book has seven chapters with high quality content covering general aspects of the fabrication method for cathode, anode, and solid electrolyte materials and their thin films. The contributors to this book have extensive experience in various facets of lithium-ion battery including materials chemistry, device fabrication, nanotechnology, etc. It is fortunate that all of the chapters have been written by experts with different backgrounds.

The development of novel devices has a strong possibility to induce innovation, because the mother of innovation, creativity, can be also produced by the unprecedented combination of apparently ordinary and well-known parameters in their field. Thus, reconstruction and coordination among the various disciplines such as physics, chemistry, materials, mechatronics, and electrics, on which energy materials are based, enhance to demonstrate the innovative thin-film devices.

This book is the product of collaborations between all contributing authors who agreed to share their research expertise and technological visions for the future. We hope this book will significantly stimulate readers to develop new devices. We would like to express our gratitude to all the authors, publishers, and other research groups. Finally, we would like to acknowledge the sincere support of Ms. Romina Skomersic and Ms. Ana Pantar of IntechOpen publishing in evolving this book.

Prof. Dr. Mitsunobu Sato and Dr. Hiroki Nagai
Kogakuin University of Technology and Engineering,
Tokyo, Japan

Dr. Li Lu
Professor,
National University of Singapore,
Singapore

Section 1

Introduction

Chapter 1

Introductory Chapter: Lithium-Ion Batteries - Thin Film for Energy Materials and Devices

Hiroki Nagai and Mitsunobu Sato

1. Thin film lithium-ion battery

In 2019, the Nobel Prize in Chemistry has been awarded to John B. Goodenough, M. Stanley Whittingham, and Akira Yoshino for their research in improving battery technology. It is the invention of lithium-ion battery (LIB). The energy density of LIB with high discharge voltage (3.6 V) is nearly twice that of Ni-Cd batteries, and excellent cycle life and higher level of intrinsic safety have been demonstrated. The LIB has revolutionized our lives and is widespread from small-scale devices such as mobile phone to emergency distributed power supply, electric vehicle, etc. Lithium-ion batteries are evolving even now. Many current types of research for LIB focus on life extension, energy density, safety, cost reduction, and charging speed.

Thin film LIB is one of the forms of LIB. It has attracted much interest for use as power sources of smart cards, implantable medical devices, micro-sensors, and so on. The thin film LIB is composed of the anode, cathode, and electrolyte with thicknesses on the order of microns. As the demands for safety, higher energy density, and other performance metrics increase, research into anode, cathode, and electrolyte materials has been rapidly progressing. Cathode materials are often mixed metal oxides involving lithium ion such as $LiCoO_2$ and $LiMn_2O_4$. Anode materials are lithium metal, carbon-based materials, and inorganic compounds. Both the cathode and anode materials are film, chosen for their ability to intercalate, and de-intercalate lithium ion while maintaining their structural integrity. The current research of electrolyte, whose form is preferable to be solid in thin film batteries, trends toward ceramics such as lithium lanthanum zinc oxide (LLZO) and lithium lanthanum titanium oxide (LLTO). The optimal electrolyte should be an efficient ion-conductor and a good electrical insulator allowing the battery to operate safely. The optimal combination of these materials can yield a battery that is light, thin, long-lasting, and safe.

2. The structure of this book

This book includes four categories: (1) method for thin film fabrication, (2) cathode, (3) anode, and (4) solid electrolyte, written by world authorities in these fields. The chapter of "Methods of Fabricating Thin Films for Energy Materials and Devices" describes the overview of the methodology for thin film fabrication. Thin film fabrication techniques can be categorized into two procedures representing dry and wet processes. Methods such as magnetron sputtering, pulsed laser, and chemical vapor deposition have been found to be capable of forming thin films of good quality and are well established across the coatings' industry. Alternative fabrication procedures such as the molecular precursor, electrospray deposition, and sol-gel method have been devised to ensure the thin film fabrication at a low cost with no complicated equipment.

Chapters 3 and 4 are related to cathode materials. The chapter 3 "Cathode Electronic Structure Impact on Lithium and Sodium Batteries Parameters "is based on her own investigations of A_xMO_2 cathode materials (A=Li, Na; M=3d) and demonstrates that the electronic structure of these materials plays an important role in the electrochemical intercalation process. The proposed electronic model of intercalation is universal and has outstanding significance with regard to tailoring the properties of electrode materials to the most efficient application in Li-ion and Na-ion batteries. The chapter reveals several correlations between electronic structure, transportation of cations, and electrochemical properties of layered Li_xCoO_2, $Li_xNi_{1-y-z}Co_yMn_zO_2$, and Na_xCoO_2 cathode materials and explains the apparently different character of the discharge/charge curve in Li_xCoO_2 and Na_xCoO_2 systems. Chapter 4 "Cathode for thin film lithium-ion batteries" describes an overview of cathode materials including lithium-containing cathode for LIB, in terms of specific capacity, energy density, working voltage, cycling life, and safety. Furthermore, some modification strategies for these cathode materials are also discussed for improving electrochemical performance. Of course, the thin film Li-ion battery applications of these cathode materials are summed up toward next-generation flexible and high-energy devices.

Chapters 5 and 6 are written regarding anode materials. The chapter 5 "Flexible Porous Carbon Nanotube Films Intercalated with Active and Functional Materials for Lithium-ion Batteries" focuses on several new carbon nanotube films for improving and enhancing the energy/power density and cyclic performance of LIB. Meanwhile, different carbon nanotube films have their own additional advantages in conductivity, strength, toughness, or thermal conductivity to meet various functional requirements of LIB. The chapter 6 "TiO_2 Nanoparticles Prepared by Sol-Gel Method for Anode Application in Lithium-Ion Batteries" describes TiO_2 electrochemical performances as anode material in LIB, which are investigated by cyclic voltammetry, galvanostatic cycles, and rate capability measurements.

The chapter 7 focuses on the solid-electrolyte interface between anode materials. The chapter 7 "Surface Chemical Analysis of Solid-electrolyte Interphase Layer on Germanium Thin Films and the Effect of Vinylene Carbonate as Electrolyte Additive" focuses on germanium thin-film anodes for LIB applications. As part of this chapter, they review briefly the use of germanium thin-films in LIB and, subsequently, new results pertaining to the effect of vinylene carbonate (VC) as electrolyte additive on the electrochemical performance are presented.

3. Conclusion

The worldwide demand for energy requires that various improved technologies for sustainable production, conversion, and storage of energy are developed. Energy materials as well as their assembled devices are urgently demanded, which are the key and foundation to realize the new energy. The reconstruction and collaboration among various disciplines of energy materials based on physics, chemistry, materials, mechatronics, and electricity are necessary to create novel thin film devices. We sincerely hope this book, which aims to develop and create innovative thin film energy devices, will be useful for the development of next-generation energy devices.

Author details

Hiroki Nagai and Mitsunobu Sato*
Kogakuin University of Technology and Engineering, Tokyo, Japan

*Address all correspondence to: lccsato@cc.kogakuin.ac.jp

Section 2

Methods of Thin Film

Chapter 2

Methods of Fabricating Thin Films for Energy Materials and Devices

Philipus N. Hishimone, Hiroki Nagai and Mitsunobu Sato

Abstract

With the continued miniaturization of the electronic devices applicable in our daily lives, thin films of various functional materials used in such devices are increasingly preferred over the traditional bulk components. Various gas-phase methods have been found to be capable of depositing thin films of good quality and are well-established across the coatings' industry. However, they are associated with ultrahigh vacuum systems and complicated and expensive instrumentation, and may involve toxic or corrosive chemical precursors. Alternative fabrication methods such as the electrospray deposition, the sol-gel method, and the molecular precursor method have been devised and represent active research areas. The molecular precursor method is relatively new. However, it has been found to be capable of effectively fabricating thin films of various metal oxides and of metals. In this chapter, some methods employed in the fabrication of the thin films are discussed in detail. The ease of practical application and relative cost-effectiveness associated with each method, the quality, and type of the fabricated thin films are also discussed. Based on the recent results by the present authors, the fabrication and characterization of a highly conductive and well-adhered thin film of metallic copper by using the molecular precursor method are presented.

Keywords: thin films, active materials, efficient electrical conductors, gas phase processes, liquid phase processes

1. Introduction

The use of highly functionalized thin films in various electronic devices has made life comfortable [1] and this is due to the enhanced functional properties of materials at the nano-scale level. At present, the miniaturization of various electronic devices is inevitable as the electronics industry looks at manufacturing thinner and lighter devices [2], in addition to reduced power consumption. For example, in secondary lithium-ion batteries (LIBs), electrodes employing thin films of active materials are preferred over their bulk counterparts due to the much improved electronic and ionic conductivity, increased specific surface area and the ease of controlling the morphology. This means that light and compact power sources with identical energy capacity to their heavier and large counterparts can be easily manufactured through thin films applications.

Thin film fabrication is not only limited to applications in LIBs. In solar cells and various electronics, for example, thin films of semiconductors and conductors are also of great importance with the aim of replacing materials that are expensive and in short supply. In this field, copper (Cu) has emerged as an ideal candidate for

IntechOpen

applications in microelectronic devices. It is abundant, and an excellent electrical conductor with a very low electrical resistivity (1.67 μΩ cm), only second to silver, but more affordable. Additionally, Cu has excellent resistance against electromigration, in comparison to aluminum which is currently used for interconnects in microelectronic devices [3].

Owing to the above-mentioned benefits of thin films' applications in LIBs and electrical devices, a considerable number of techniques have been established for the fabrication of these thin films. In this chapter, some of the effective techniques used in the deposition of thin films are described and discussed in Section 2.

2. Established methods for fabricating thin films

The fabrication techniques can be divided into two categories representing (1) gas phase and (2) liquid phase fabrication methods for thin films as outlined in **Table 1**. In this context, the gas phase fabrication methods are methods that involve the deposition of materials either from small particles of bulk solid materials or chemical precursors in vapor form whereas the liquid phase fabrication methods refer to the methods involving a reaction between a substrate and a chemical precursor in a liquid state.

1. Gas phase fabrication methods	2. Liquid phase fabrication methods
1. Magnetron sputtering	1. Electrospray deposition
2. Pulsed laser deposition	2. Electrochemical deposition
3. Atomic layer deposition	3. Sol-gel method
4. Chemical vapor deposition	4. Molecular precursor method

Table 1.
List of some of the methods for fabricating thin films for LIBs and devices.

2.1 Gas phase fabrication methods

2.1.1 Magnetron sputtering

This is one of the most well-established techniques that is widely used in the industries to fabricate coatings of many different materials, including metals, semiconductors, and alloys [4, 5]. Thin films fabricated *via* magnetron sputtering prove to be of superior quality in comparison to those fabricated using other gas phase methods [4, 6]. The schematic representation of the magnetron sputtering set-up is shown in **Figure 1**.

Before deposition, the chamber is evacuated to ca. 10^{-4} kPa and then refilled with Ar gas which is used as the sputter gas. Using strong electrical and magnetic field, magnetron sputtering employs the bombardment of the selected target's surface by Ar^+ ions which are formed as a result of the collision between Ar gas atoms and electrons trapped within the magnetic field. The Ar^+ ions are attracted toward the negatively charged target (cathode) and the bombardment leads to the ejection of the targets particles which are then redeposited as a thin film on the substrate placed on the anode, opposite to the target. Depending on the type of power source utilized, magnetron sputtering can be divided into two types namely, (1) direct current (dc) magnetron sputtering and (2) radio frequency (rf) sputtering. While dc magnetron sputtering is relatively cheaper in comparison to rf, only electrically conductive targets like metals or doped semiconductors are used [7]. Additionally,

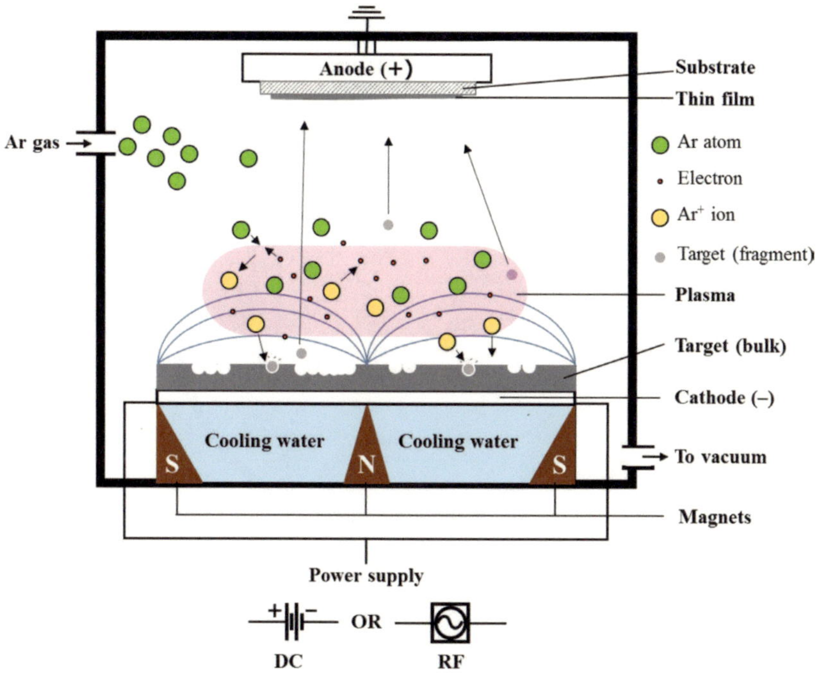

Figure 1.
Schematic representation of a magnetron sputtering equipment and deposition process.

reactive gases such as oxygen or nitrogen gas are used to react with the sputtered materials, in the fabrication of thin films for metal oxides or nitrides, respectively. The technique is referred to as "reactive magnetron sputtering". **Table 2** lists the superiorities and limitations associated with the magnetron sputtering technique.

Superiorities	Limitations
1. Well-established for the deposition of various elements, alloys, and compounds 2. Ability to fabricate thin films of high quality and identical composition with the target materials	1. Expensive, complicated and ultrahigh-vacuum systems required 2. Low deposition rates and low ionization efficiencies 3. Expensive target materials. 4. High substrate effect by the sputtered particles

Table 2.
Superiorities and limitations associated with the magnetron sputtering technique.

2.1.2 Pulsed laser deposition (PLD)

The schematic representation of a PLD setup is shown in **Figure 2**.

The PLD technique employs the ablation of a target material with a strong, pulsed laser beam to produce a plume of vaporized materials which is then recondensed and deposited onto a substrate, placed opposite the target, under a reduced pressure atmosphere of ca. 10^{-1}–10^{-5} kPa [8–10]. Depending on the target material, its morphology, and the laser pulse wavelength and duration, there is a specific threshold power density that is required to cause ablation [11]. Therefore, a good understanding of the various parameters is required for a successful plume generation and subsequent thin film deposition. The stoichiometric transfer of

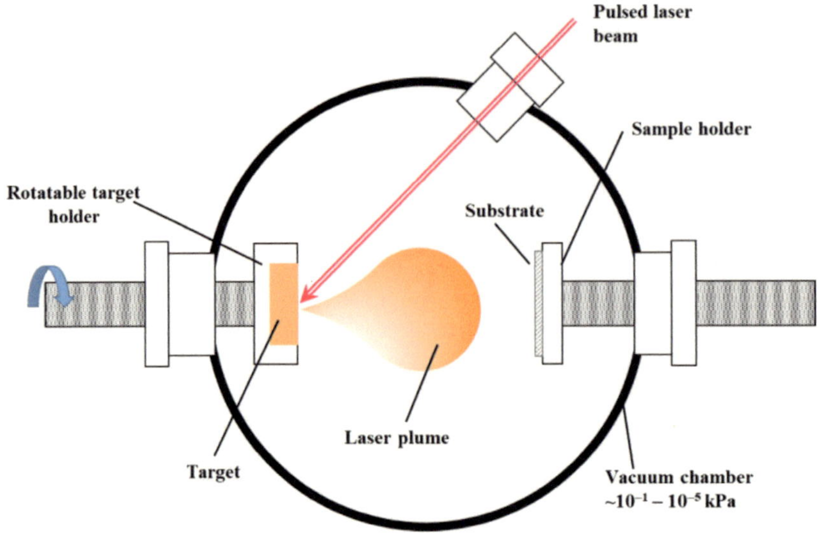

Figure 2.
Schematic representation of PLD setup.

materials from the target to the substrate earned PLD a reputation as fabrication of choice for thin films with complicated stoichiometry or sandwich structures [12], in comparison to other gas phase techniques.

Although the fabrication of thin films with identical stoichiometry as the target materials is one of the superior characteristics of PLD, the stoichiometry can be compromised when multi-component materials with volatile constituents are used [11–13]. As a remedy, a background gas such as oxygen or nitrogen is introduced in the reaction chamber to compensate for the lack of stoichiometry, during the fabrication of metal oxides or nitrides' thin films, respectively. The superiorities and limitations of the PLD technique are summarized in **Table 3**.

Superiorities	Limitations
1. Stoichiometric transfer of materials from the target to the thin film 2. Ideal for fabricating multi-layered thin films 3. Ideal for the deposition of thin films for high-T_c superconductors	1. Expensive, complicated and ultrahigh-vacuum systems are required 2. Difficult to scale-up for mass production 3. Unwanted ablation impurities are often incorporated in the thin film

Table 3.
Superiorities and limitations associated with the PLD technique.

2.1.3 Atomic layer deposition (ALD)

The ALD technique is based on sequential and self-limiting reactions of a chemical precursor in vapor form, with an activated or functionalized surface of the substrate [14, 15]. The self-limiting property of this technique is realized in a sense that, if all functional sites on the substrate have reacted, no further reaction between the chemical precursor and the substrate will take place [14, 16]. **Figure 3** illustrates the concept of ALD in the fabrication of a TiO2 thin film. The steps are defined as follows: (a) precursor exposure, (b) purge, (c) reactant exposure, and (d) purge. By repetition of step (a)–(d), layers of TiO_2 can be easily formed and the thickness can be controlled at the atomic level. Under a vacuum pressure of ca. 10 kPa or less, the chemical precursors are maintained in a vapor phase by adjusting a specific temperature referred to

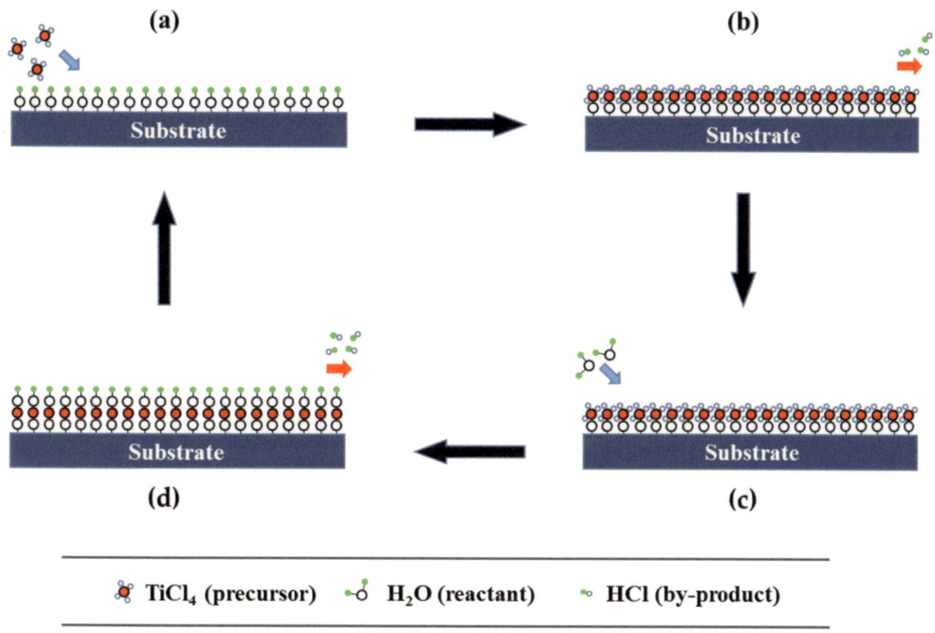

Figure 3.
Schematic representation of 1 cycle of an ALD process for the deposition of a TiO_2 thin film from titanium tetrachloride and water.

as the "ALD temperature window" [17], to achieve optimum reaction kinetics and avoiding condensation or thermal decomposition of the precursor.

The superiority of ALD over other gas phase processes for the deposition of thin films is realized through its exceptional conformity, the ability to control thickness at the atomic level and tunable film composition [15, 18, 19]. Through this route, the deposition of thin films on substrates with highly-structured geometries can be easily achieved, and ALD has been considered as the ideal method for fabricating highly conformal seed layers [20, 21]. This is also of great importance for applications in the concept of integrated batteries [22], that aims at maintaining the storage capacity of miniaturized LIBs. Although the fabrication of metallic copper thin films has been achieved [17, 21, 23–25], the deposition of metallic thin films is generally limited due to the lack of precursors suitable for ALD processes. The vaporization of chemical precursors, difficulty in controlling their thermal decomposition and lack of efficient self-limited reactivity with surfaces are some of the challenging aspects in ALD [19]. In addition, during the designing stages of certain chemical precursors, the by-products associated with the proposed chemical reaction need to be carefully considered as toxic or corrosive chemicals are likely to be those by-products. **Table 4** lists the superiorities and limitations of the ALD technique.

Superiorities	Limitations
1. Suitable for the fabrication of highly conformal thin films 2. Chemical reactions are highly selective toward the substrates and are self-limiting 3. Thickness control at atomic scale allows for the deposition of ultra-thin films	1. Low deposition rates 2. Requires functionalized substrates to ensure reaction with the chemical precursors 3. The technique is limited due to the unavailability of suitable chemical precursors

Table 4.
Superiorities and limitations associated with the ALD technique.

2.1.4 Chemical vapor deposition (CVD)

Defined as the deposition of solids onto heated substrates from chemical reactions in vapor phase [26, 27], CVD represents a versatile deposition technique of thin films for a wide range of materials, under vacuumed atmospheres and temperatures over 600°C, and it is an established technique for the deposition of thin film for microelectronics, ohmic and Schottky diode contacts, diffusion barriers, wear-resistant coatings, etc. [26–29]. The schematic representation of a CVD setup is shown in **Figure 4**.

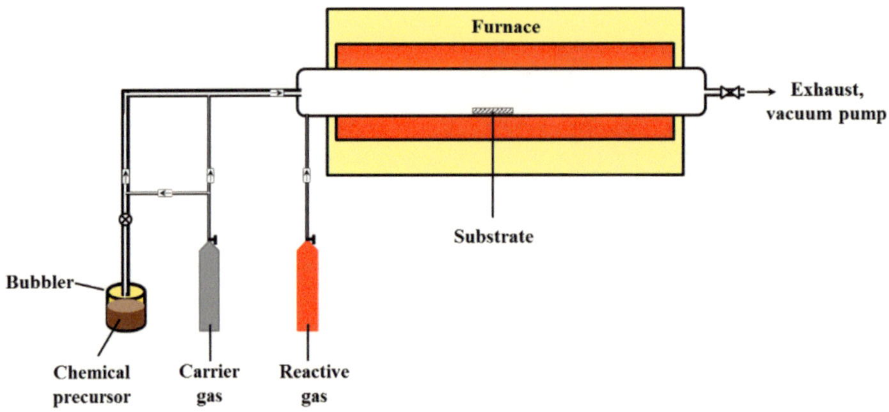

Figure 4.
Schematic representation of a CVD setup.

In a typical CVD experiment, a combination of chemical precursors showing stable vaporization behavior is carefully selected. Using an inert carrier gas, usually Ar or N_2 gas, the vaporized precursors are carried into an evacuated furnace (reaction chamber) and mixed with an additional reactive gas or two. In the fabrication of metal oxide thin films, oxygen gas is used as the additional carrier/precursor gas. On the other hand, if the desired thin film is metallic, such as Cu, a reducing atmosphere is created by using H_2 gas [30, 31].

The CVD technique benefits from the fact that it does not require ultrahigh vacuum systems like in the magnetron sputtering and PLD techniques [26, 32]. In addition, CVD is a kinetically driven process, associated with minimized agglomeration and also benefiting from faster deposition rates, in comparison to ALD [32, 33]. Because different precursors have different evaporation rates, it is generally difficult to deposit multicomponent thin films by CVD. Other drawbacks of CVD include possible chemical and safety hazards of toxic, corrosive and flammable precursors or exhaust gases, and poor thin film adhesion-strength [27, 28]. The superiorities and limitations associated with the CVD technique are summarized in (**Table 5**).

Superiorities	Limitations
1. High deposition rates 2. High throwing power that allows for the coating of deep recesses, holes and other complicated 3-D configurations 3. Does not require ultrahigh-vacuum systems	1. High temperatures are required to decompose the chemical precursors 2. Chemical precursors with high vapor pressure are often hazardous and toxic 3. The required neutralization of toxic and corrosive byproducts raises the cost of the technique

Table 5.
Superiorities and limitations associated with the CVD technique.

2.2 Liquid phase processes

2.2.1 Electrospray deposition (ESD)

ESD involves the deposition of charged particles formed as a result of liquid atomization by an electrical force. The atomization of the liquid by electrical force allows for the generation of small particles, of narrow size distribution which can be uniformly dispersed over the substrate and, ESD has been considered as a promising technique for the formation of high-quality layers and films [34]. If the ESD process is carried out at elevated temperatures, thin film formation is achieved through the thermal decomposition of the precursor solution during deposition [35] and the process is referred called "electrospray pyrolysis (ESP)." The schematic representation of the ESD setup is shown in **Figure 5**.

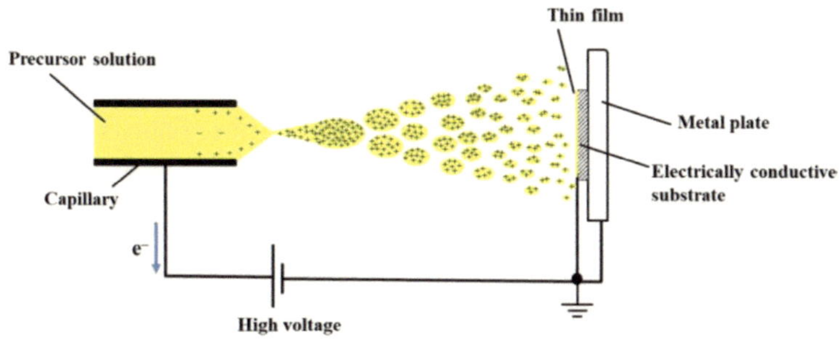

Figure 5.
Schematic representation of an ESD setup.

As shown in **Figure 5**, the ESD method is based on supplying a precursor solution through a metallic capillary which is maintained at high electrical potential. Ideally, the solution forms a Taylor cone at the tip of the capillary and emitted as a solution jet composed of many highly charged droplets that are homogeneously dispersed in the space between the nozzle and the counter metallic plate (substrate holder) by Coulomb repulsion forces [34]. The quality of the resultant thin films can be easily controlled by adjusting variables such as the applied voltage, distance between the atomizing nozzle and, the flow rate, concentration and amount of the precursor solution.

In addition to a very low throughput associated with the ESD technique, thin film deposition on insulating substrates is generally not possible due to charge accumulation on the substrate, which leads to the repulsion of more incoming charged particles [34]. Modifications such as the use of alternating current (ac) [36] and extractor electrodes [34] have been employed to solve the charge repulsion issue and ensure the thin film deposition on insulating substrates *via* ESD. However, this complicates an initially simple experimental setup. The superiorities and limitations of the ESD technique are listed in **Table 6**.

Superiorities	Limitations
1. Simple instrumentation 2. Uniform coating on large areas and easy controlling of the surface morphology 3. The rate of deposition can be easily controlled by adjusting the voltage and flow rate	1. The strong electric field is required 2. Thin film fabrication on insulating substrates is challenging 3. The technique is associated with low throughput 4. The technique is very sensitive to the physical properties of the used liquid and it's very difficult to obtain a stable cone-jet throughout the experiment

Table 6.
Superiorities and limitations associated with the ESD technique.

2.2.2 Electrochemical deposition (ECD)

Considered as one of the cheapest methods for the deposition of thin films, ECD represents a major technology for mass production of large area metallic protective coatings in the industry [37]. The deposition technique employs a combination of reduction and oxidation reactions of chemical precursors in electrolytes of aqueous solutions, organic solutions or fused salts [38, 39]. In addition to cost, other advantages of the ECD include low fabrication temperature, high purity and, the ease of controlling thin film thickness, uniformity and rate of deposition by adjusting the deposition parameters [40–42]. The electrochemical technique can be divided into two groups, namely, (1) electrolysis plating which uses external power source to drive the electrochemical reaction that leads to the deposition of a thin film onto an electrically conductive substrate and, (2) electroless plating which does not require external power supply but requires the use of catalysts to activate the surface of the specimen. The schematic representations of the electrochemical techniques for the fabrication of metallic copper thin films are given in **Figure 6**.

During electrolysis deposition, the external power source causes the dissolution of the metal from the anode (usually a copper rod or plate). The dissolved Cu^{2+} ions are then reduced and redeposited on the surface of the electrically conductive substrate (Cathode). An electrolytic solution involving a salt of copper such as $CuSO_4$ is used for the electrochemical bath to maintain the balance of Cu^{2+} ions. In electroless plating, the electrolytic solution is also the source of the Cu^{2+} ions and a reducing agent such as formaldehyde (HCOH) is added to the plating bath. The introduction of a substrate with an activated surface (by $PdCl_2$ or $SnCl_2$) initiates the reduction of Cu^{2+} to metallic copper, subsequently forming a film on the surface of the substrate. In both the electrolytic and electroless plating, various alkaline or acidic electrolytic solutions are used and are important for maintaining the pH and stabilities of the plating baths [43].

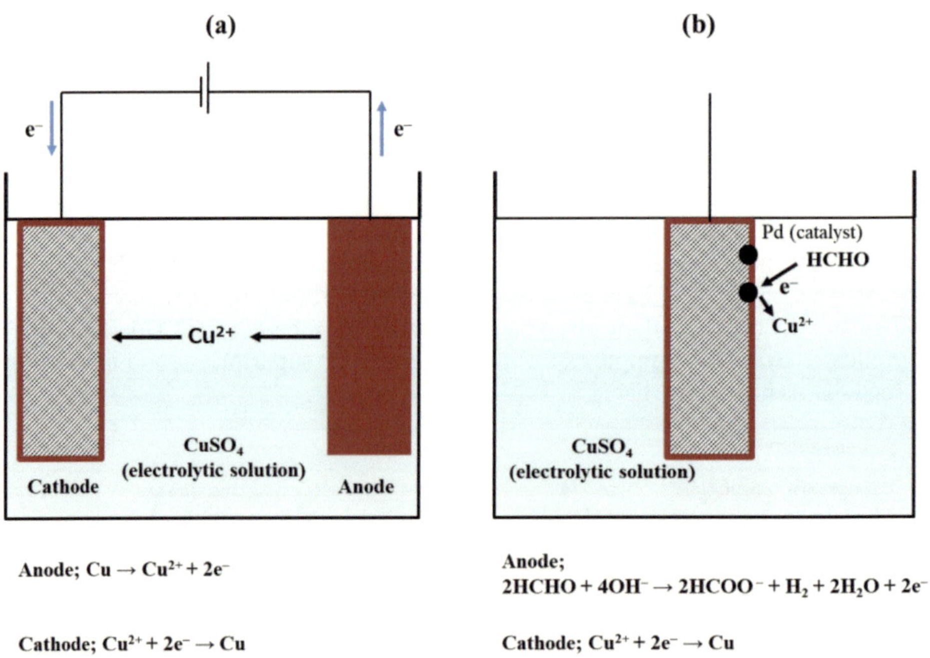

Figure 6.
Schematic representation of the electrochemical deposition techniques. (a) Electroplating. (b) Electroless plating.

Unlike in the field of metal plating, the deposition of thin films with semiconducting properties is challenging due to complications in controlling the reaction kinetics of multiple compounds in the solutions, at ppm level [37]. However, the major breakthrough came with the preparation of CdTe layers for solar cells and modules and, since then films of semiconducting materials such as CIGS [37] and $LiMn_2O_4$ [44] have been successfully fabricated. Although electroless plating is considered as a promising technique for the deposition of thin films especially on insulating substrates, the use of $PdCl_2$ or $SnCl_2$ compromises the quality of the thin films [45]. In addition, the selection of the chemical components for the electrolytic solution is important for ensuring that the reduction reaction occurs only on the surface of the activated substrate [43]. **Table 7** lists some of the well-known superiorities and limitations of the ECD techniques.

Superiorities	Limitations
1. Inexpensive instrumentation 2. Applied at room temperatures and pressures 3. Properties of the thin films can be easily controlled by modifying the electrochemical variables 4. Suitable for large area deposition	1. Poor adhesion, the inclusion of cracks, blisters, and pinholes in the resultant thin films 2. Toxic supporting electrolytes or complexing agents such as cyanide are used 3. The substrate must conductive. If not, a catalyst is required

Table 7.
Superiorities and limitations associated with the ECD technique.

2.2.3 Sol-gel method

The sol-gel method has emerged as a method of choice for the fabrication of ceramics and glasses and, it is considered as a cost-effective alternative to the well-established gas phase processes. The sol-gel method is a two-chemical processes technique based on the hydrolysis and condensation of molecular precursors and has since been considered to be capable of offering new routes for the low-temperature fabrication of oxide materials [46]. Because all gel products may contain nanoparticles or are nano-composites, the sol-gel method has been deemed as typical nanotechnology [47] and the fabrication of thin films can be easily achieved after its integration with spin-coating or dip-coating.

In a typical sol-gel process, metal precursors (of the desired metal oxide) derived from metal salts in aqueous solutions, metal alkoxides in organic solvents, or a combination of the two are used. During the hydrolysis step, metal-hydroxide groups are formed by changing the pH of the aqueous solution or adding water to the metal alkoxides. The subsequent condensation reaction results in the formation of metal-oxygen-metal bonds, with different polymerization and poly-condensation processes leading to the gradual formation of the solid phase network [48]. Although a network of metal-oxygen-metal bonds of the desired metal oxide is obtained from the polycondensation step, additional drying and annealing steps are required for the removal of organic materials and improvement of the crystallinity of the resultant thin film. A typical sol-gel process for the formation of SiO_2 from silicone alkoxides is given in **Figure 7** [49].

Although the sol-gel method has become an indispensable technique in the fabrication of thin films for metal oxides, principally, the fabrication of metallic thin films is not viable. In addition, a good understanding is required for the selection of appropriate precursors and the sophisticated chemical processes leading to the hydrolysis and condensation. The stabilization of the transition metals against hydrolysis, condensation, and other nucleophilic attacks is generally challenging. It

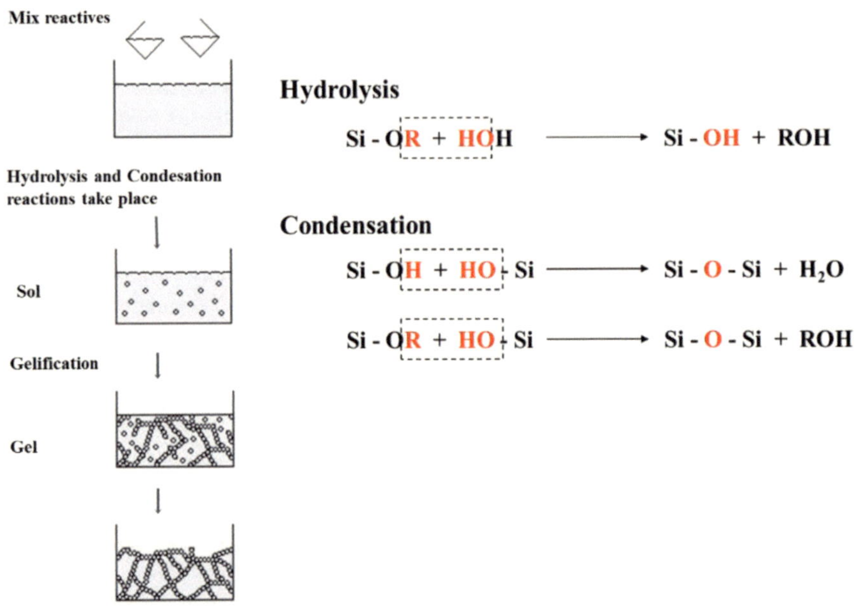

Figure 7.
Typical sol-gel process for SiO_2 formation from silicon alkoxides [49].

is also important to make sure that the transition metal is sufficiently saturated in solution, to avoid different reaction mechanism. Finally, proper control of the moisture and hydrolysis conditions is required to avoid precipitation of the transition metal alkoxides [48]. Therefore, all these aspects make the preparation of suitable coating precursors for the sol-gel method quite complicated. The superiorities and limitations of the sol-gel method are presented in **Table 8**.

Superiorities	Limitations
1. Inexpensive instrumentation 2. Thin films of metal oxides can be fabricated at relatively low temperatures in comparison to the conventional melting technique	1. The chemical precursors are not stable against hydrolysis and side reactions 2. High annealing temperatures are required to improve the crystallinity of the film due to the involved rearrangement of the metal-oxygen-metal network

Table 8.
Superiorities and limitations associated with the sol-gel technique.

2.2.4 Molecular precursor method (MPM)

The molecular precursor method is an effective, wet chemical process that was developed by the current authors, for fabricating nano-crystalline thin films of metals and, various metal oxides and phosphates [1, 49–53]. In the fabrication of these thin films by the MPM, metal complexes are dissolved in ethanol by combining them with appropriate alkylamines and this yields good coating solutions, with many practical advantages and are ideal for various coating procedures such as spin-coating, dip-coating, and flow-coating. Additionally, the development of aqueous precursor solutions have been successfully achieved [54, 55] and, these solutions can be easily applied for the fabrication of thin films *via* spray-coating at ambient conditions. To date, over 40 different kinds of thin films for metal oxides or phosphates have been easily fabricated by using the MPM. The procedure employed by

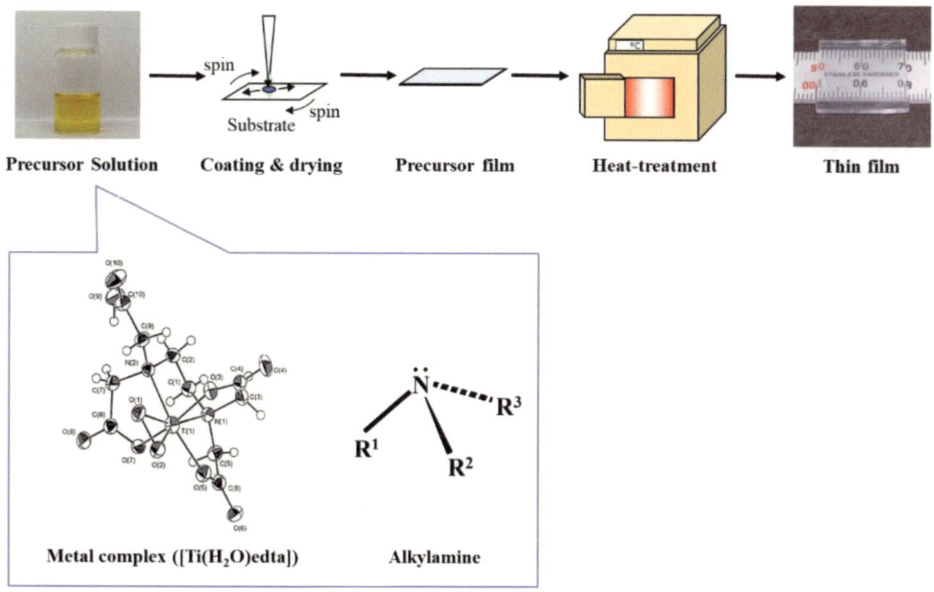

Figure 8.
Procedure for fabricating a titania thin film via *the MPM.*

the MPM is represented by the example of fabricating the titanium dioxide thin film [1] illustrated in **Figure 8**. The procedure starts with the preparation of a coating solution by reacting an isolated Ti(IV) complex of ethylenediamine-*N,N,N′,N′*-tetraacetic acid (EDTA) with dipropylamine in ethanol. The precursor solution, which shows stability against up to 10% water [56], is suitable for spin-coating onto a glass substrate and dried at 70°C to yield an amorphous film. In order to obtain a thin film of metal oxide spread homogeneously on the substrate after the subsequent heat-treatment, it is important that the precursor film is amorphous. Just as the case with the metal/organic polymers in the sol-gel method. The precursor film is then heat-treated at 450°C in air, to eliminate the organic components involved and a transparent thin film of titania is obtained.

In comparison to the sol-gel method, the MPM produces precursor solutions that are stable and can be stored for longer periods without being hydrolyzed. It is considered that thin films with much smaller crystallite sizes can be fabricated by the MPM due to the ease of metal oxides' nucleation. This is a result of a large number of crystallites instantly formed during the removal of organic ligands *via* the heat-treatment step of the MPM, in comparison to the rearrangement of polymer chains during the same stage in the sol-gel method [49]. Consequently, thin films of metal oxides can be easily fabricated at relatively low annealing temperatures *via* the MPM in comparison to the sol-gel method.

By following the procedure outlined in **Figure 8**, the fabrication of thin films for a novel thin-film lithium-ion battery (LIB) which can be charged by light irradiation, and thus functioning as a photovoltaic-LIB was recently achieved [57]. The device uses thin films of TiO_2 and $LiCoO_2$ deposited on fluorinated tin oxide (FTO) pre-coated glass substrates, as the anodic and cathodic active materials, respectively, and an electrolytic solution containing $LiPF_6$. In another study employing the MPM, the fabrication of the first *p*-type Cu_2O transparent thin film with properties identical to those of thin films fabricated by using gas phase processes was also achieved [58]. A dry-type solar cell utilizing this Cu_2O transparent thin film combined with a Vis-responsive anatase thin film also fabricated by the MPM could be successfully constructed and evaluated [49].

2.2.5 *Highly conductive and well-adhered Cu thin film fabricated by heat treatment of a precursor film obtained by spray-coating an aqueous solution involving Cu(II) complexes*

The compatibility of the MPM with the spin-coating and dip-coating techniques in the fabrication of a Cu thin film and embedding Cu in Si trenches has been described in our earlier reports [59, 60]. Very recently, our group succeeded in the fabrication of a highly conductive and well-adhered thin film of copper by heat-treatment of a precursor film obtained via spray-coating [55]. Spray-coating and other spray-on related techniques benefit from the relatively simple and inexpensive instrumentation set-up, reduced material losses [61], ability to be adjusted for large area deposition and produce thin films in a Roll-to-Roll and high throughput process in comparison to other solution-based techniques such as spin coating [62].

It is however important that a precursor solution suitable for spray-coating is developed. In our recent study, the stable and VOC-free solution was developed by mixing a Cu(II) complex of EDTA ([Cu(H_2edta)]•H_2O) and Cu(II) formate with ammonia, in water. The ratio of [Cu(H_2edta)]:Cu(II) formate was 1:4 and the total concentration of Cu^{2+} in the precursor solution was 0.1 mmol g^{-1}. The aqueous solution could be easily sprayed onto a quartz glass substrate pre-heated to 180°C, without forming solids that may clog the nozzle tip of the airbrush. The sprayed film denoted as **Fas-sprayed** was heat-treated at 350°C for 50 min in Ar gas (the resulting film is denoted **FHT**), followed by a post-annealing step whereby an identical-sized, clean quartz glass was placed on top and heat-treating at 400°C for 50 min also in Ar gas. The final resultant thin film is denoted as **FHP**. The XRD patterns of **Fas-sprayed, FHT,** and **FHP** are shown in **Figure 9**.

The XRD pattern of the as-sprayed film shows peaks assignable to the (111), (200) and (220) crystal phase of metallic copper at 43.4, 50.5 and 74.2°, respectively (ICDD card no. 00-004-0836). The pre-heating temperature was sufficient to induce the crystallization of metallic copper. However, although the spray-coating was carried out in the air, at a substrate temperature of 180°C, the crystallized Cu^0 was not oxidized as there were no peaks assignable to the oxide species of copper observed in the XRD spectrum of **Fas-sprayed.** It is proposed that the carbon atoms derived from the EDTA ligand could effectively protect the Cu^0 against oxidation

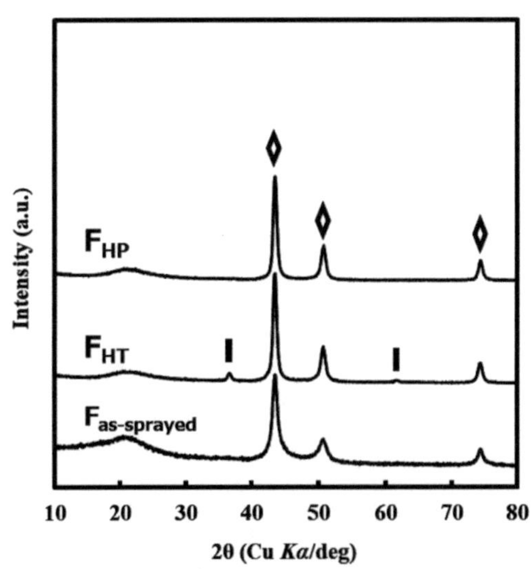

Figure 9.
XRD patterns of **Fas-sprayed, FHT,** *and* **FHP***. The peaks are denoted as follows:* ◊ *Cu and* ▮ *Cu_2O.*

at a preheating temperature of 180°C. After the heat-treating procedure, the XRD pattern of **FHT** shows additional peaks at 36.6° and 61.6° assignable to the (111) and (220) crystal phase of Cu_2O, respectively (ICDD card no. 01-071-3645). It is however notable that the as-sprayed film was not extensively oxidized by heat-treating at 350°C for 50 min under an Ar gas atmosphere containing about 10 ppm of air as an impurity. The Cu_2O phase was successfully removed during the post-annealing procedure whereby the carbon atoms remaining within the film were used to reduce the Cu^+ to Cu^0. As a result, the XRD pattern of the **FHP** film shows only peaks assignable to metallic copper.

Figure 10 shows the Auger Electron Spectroscopy (AES) measurements of **FHP**. The surface scan (**Figure 10a**) indicated that the thin film is composed of C and Cu, as has been confirmed by one peak at 265 eV assignable to electrons of carbon atom and three peaks (LVV) at 764, 835, and 914 eV assignable to electrons of Cu, respectively. The curve-fitting result (**Figure 10b**) revealed that all carbon atom present in the thin film is neutral. The effectiveness of the carbon atoms present in the film has been illustrated in the protection of the **FHT** film against excessive oxidation. Specifically, the ratio of Cu_2O:Cu peaks assignable to the (111) plane of each phase in **FHT** was 4:96. This implies that the carbon atoms derived from the EDTA ligand and formic acid [of Cu(II) formate] were vital for further protection of the formed Cu^0 phase in an Ar gas atmosphere containing less than 2 ppm of oxygen gas. On the basis of our previous procedures, the post-annealing procedure accounts for the removal of up to 50% of the carbon atoms present in the film, right after the heat-treatment procedure. Therefore, it is acceptable that the neutral carbon is existing in the resultant thin film, **FHP**. Interestingly, the carbon atoms did not disrupt the electrical conductivity of the film.

Figure 11 shows the FE-SEM (a) and AFM (b) images of **FHP**. The images show morphologies composed of well-connected Cu grains and a densely deposited matrix of Cu grains can be observed from the FE-SE cross-sectional image, consequently promoting electrical conductivity within the film. The crack-free and densified film **FHP** of 100 nm thickness has an electrical resistivity of $3.8(6) \times 10^{-5}\ \Omega$ cm.

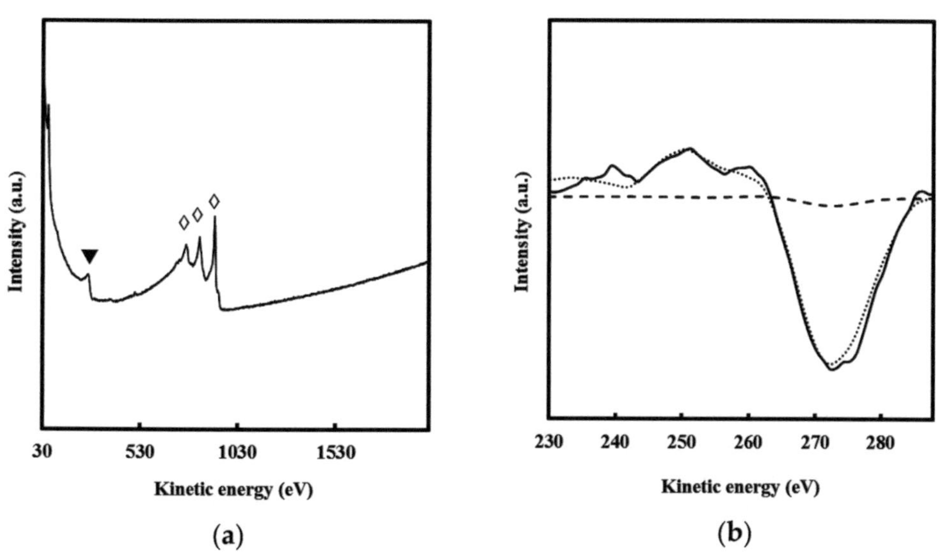

Figure 10.
(a) Auger spectrum of **FHP***. Peaks are denoted as follows: ◊ Cu and ▼ C. (b) Curve-fitting results for the carbon state in* **FHP***. The solid, dotted, and dashed lines represent* **FHP***, C (neutral) and C (tetravalent), respectively.*

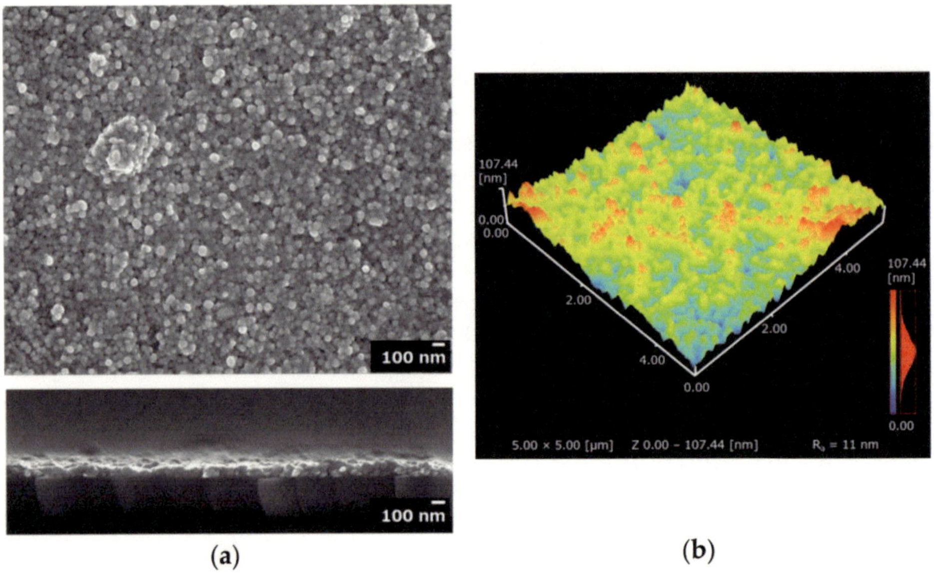

Figure 11.
(a) FE-SEM images showing the top and cross-section views of the **FHP** *film. (b) 3D AFM image of the* **FHP** *film.*

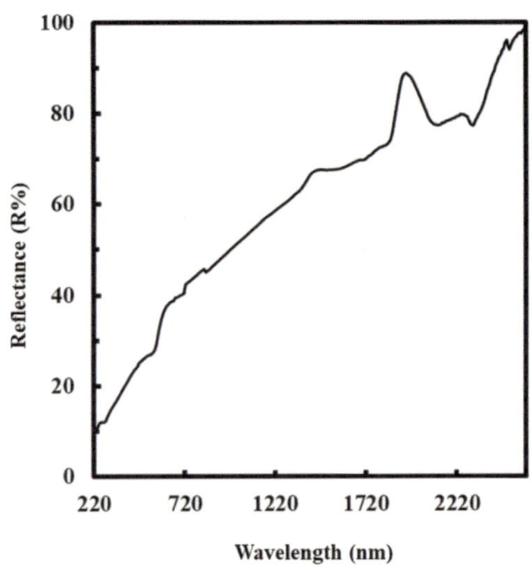

Figure 12.
Reflectance spectra of **FHP**.

Figure 12 shows the reflectance spectra of the **FHP** film. The reflectance of **FHP** approaches over 90% in the far infrared region which is comparable to the conductive and semi-transparent Cu thin film previously reported by our group [59] and can also be useful as an excellent reflector in the far-infrared region. Results of the stud-pull coating adhesion test indicated that the thin film strongly adhered onto the quartz glass substrate with an adhesion strength of up to 37 MPa. This adhesion strength was efficiently attained without any chemical or physical modification of the substrate's surface and it is expected to be a result of the interface having bonds between Cu and the O^{2-} belonging to the quartz glass substrate.

The research for the next generations is looking at the utilization of environmentally friendly reagents to achieve the fabrication of functional thin films for various materials. Therefore, the aqueous precursor solution used in our recent study has the potential as a promising technique for the fabrication of metallic Cu thin films for various applications. In our study, solutions involving different Cu(II) complexes were investigated. For example, a nonconductive film of copper was obtained after heat-treating a film from a solution involving the Cu(II) complex only. In addition, it is important to point out that the precursor solution with a [Cu(H_2edta)]:Cu(II) formate ratio of 1:4 gave the best results with good reproducibility. The importance of designing and mixing different Cu(II) complexes involved in the coating solution was clearly illustrated. By varying the ratio of the Cu(II) complexes in the coating solution, films with different crystal structures, morphologies, and electrical properties could be obtained. And this is the advantage of the MPM which enables to design metal complexes in coating solutions, at the molecular level. By using the spray-coating technique, our group also managed to fabricate a thin film of LCO on a non-crystalline quartz glass substrate.

Established just over 22 years ago, the MPM is a relatively new method in comparison to other thin film fabrication methods described earlier in this chapter. However, it is emerging as an effective chemical method for the fabrication of various transparent thin films of metal oxides and highly conductive thin films of metallic Cu. The superiorities and limitations of the MPM are given in **Table 9**.

Superiorities	Limitations
1. Stable and homogeneous precursors of different kinds of metal complexes can be easily prepared 2. The coating solutions can be tailored for different coating techniques 3. The fabrication of metal oxide thin films can be achieved at lower annealing temperatures as compared to other liquid phase processes such as the sol-gel method	1. Thermal decomposition and removal of the organic components from the precursor films are necessary

Table 9.
Superiorities and limitations associated with the MPM technique.

3. Conclusions

Various fabrication techniques for thin films applicable to energy materials and devices have been described and discussed in details. Gas phase processes such as magnetron sputtering and PLD have been identified to be capable of fabricating thin films of high quality and are well-established in the coating industry. However, expensive and complicated instruments are required. The ALD and CVD techniques are suitable for the fabrication of thin films onto substrates with complicated 3-D structures. However, potentially toxic and corrosive chemical precursors are used. Additionally, the fabrication of thin film with different chemical components is challenging due to the different properties of chemical species in the vapor phase.

The liquid phase processes are cost-effective alternatives to their gas phase counterparts. The ESD benefits from the ability to control the surface morphology of the deposited thin film but suffers from low throughput and difficulties to deposit films onto insulating substrates. ECD is well-established for the fabrication of thin films for metallic coatings such as copper, on a large area. However, the resultant thin films suffer from weak adhesion and poor surface morphology. The sol-gel method has been extensively developed for the fabrication of thin films for metal oxides at relatively low-costs in comparison to the gas phase processes. The poor stability of the chemical precursors against hydrolysis and deviating reaction mechanisms are some of the challenges encountered.

The MPM is an emerging liquid phase process capable of fabricating thin films of metal oxides such as TiO_2, $LiCoO_2$, and p-type Cu_2O etc. and the functionalities of these thin films in energy devices have been evaluated in a PV-LIB [53] and a dry-type solar cell [49]. Thin films of metallic copper have also been successfully fabricated by the MPM. This shows that the MPM is earning its place among affordable and effective techniques for the fabrication of thin films for future energy materials and devices.

Although there are other additional fabrication techniques, the chapter provided an insight into some of the well-established and promising techniques for the fabrication of future energy materials and devices, taking into consideration the costs in terms of the required instrumentation and thin film deposition conditions.

Conflict of interest

The authors declare no conflict of interest.

Author details

Philipus N. Hishimone[1,2], Hiroki Nagai[3] and Mitsunobu Sato[3*]

1 Applied Chemistry and Chemical Engineering Program, Graduate School, Kogakuin University, Tokyo, Japan

2 Department of Chemistry and Biochemistry, Faculty of Science, University of Namibia, Mandume Ndemufayo Avenue, Windhoek, Namibia

3 Department of Applied Physics, School of Advanced Engineering, Kogakuin University, Tokyo, Japan

*Address all correspondence to: lccsato@cc.kogakuin.ac.jp

References

[1] Nagai H, Sato M. Molecular precursor method for fabricating *p*-type Cu_2O and metallic Cu thin films. In: Nikitenkov N, editor. Modern Technologies for Creating the Thin-film Systems and Coatings. Rijeka: InTech; 2017. pp. 3-20. DOI: 10.5772/66476

[2] Bourzac K. Japan gambles on displays. Nature. 2012;**484**:301. DOI: 10.1038/484301a

[3] Murarka SP, Hymes SW. Copper metallization for ulsi and beyond. Critical Reviews in Solid State and Materials Sciences. 1995;**20**:87-124. DOI: 10.1080/10408439508243732

[4] Kelly P, Arnell RD. Magnetron sputtering: A review of recent developments and applications. Vacuum. 2000;**56**:159-172. DOI: 10.1016/S0042-207X(99)00189-X

[5] Swann S. Magnetron sputtering. Physics in Technology. 1988;**19**:67-75. DOI: 10.1088/0305-4624/19/2/304

[6] Bräuer G, Szyszka B, Vergöhl M, Bandorf R. Magnetron sputtering - milestones of 30 years. Vacuum. 2010;**84**:1354-1359. DOI: 10.1016/j.vacuum.2009.12.014

[7] Hahn H, Averback RS. The production of nanocrystalline powders by magnetron sputtering. Journal of Applied Physics. 1990;**67**:1113-1115. DOI: 10.1063/1.345798

[8] Lowndes DH, Geohegan DB, Puretzky AA, Norton DP, Rouleau CM. Synthesis of novel thin-film materials by pulsed laser deposition. Science. 1996;**273**: 898-903. DOI: 10.1126/science.273.5277.898

[9] Smith HM, Turner AF. Vacuum deposited thin films using a ruby laser. Applied Optics. 1965;**4**:147. DOI: 10.1364/AO.4.000147

[10] Singh RK, Narayan J. Pulsed-laser evaporation technique for deposition of thin films: Physics and theoretical model. Physical Review B. 1990;**41**:8843-8859. DOI: 10.1103/PhysRevB.41.8843

[11] Willmott PR, Huber JR. Pulsed laser vaporization and deposition. Reviews of Modern Physics. 2000;**72**:315-328. DOI: 10.1103/RevModPhys.72.315

[12] Schou J. Physical aspects of the pulsed laser deposition technique: The stoichiometric transfer of material from target to film. Applied Surface Science. 2009;**255**:5191-5198. DOI: 10.1016/j.apsusc.2008.10.101

[13] Ashfold MNR, Claeyssens F, Fuge GM, Henley SJ. Pulsed laser ablation and deposition of thin films. Chemical Society Reviews. 2004;**33**: 23-31. DOI: 10.1039/b207644f

[14] Lu J, Elam JW, Stair PC. Atomic layer deposition - sequential self-limiting surface reactions for advanced catalyst “bottom-up” synthesis. Surface Science Reports. 2016;**71**:410-472. DOI: 10.1016/j.surfrep.2016.03.003

[15] Johnson RW, Hultqvist A, Bent SF. A brief review of atomic layer deposition: From fundamentals to applications. Materials Today. 2014;**17**:236-246. DOI: 10.1016/j.mattod.2014.04.026

[16] George SM. Atomic layer deposition: An overview. Chemical Reviews. 2010;**110**:111-131. DOI: 10.1021/cr900056b

[17] Knisley TJ, Kalutarage LC, Winter CH. Precursors and chemistry for the atomic layer deposition of metallic first row transition metal films.

Coordination Chemistry Reviews. 2013;**257**:3222-3231. DOI: 10.1016/j.ccr.2013.03.019

[18] Leskelä M, Ritala M. Atomic layer deposition chemistry: Recent developments and future challenges. Angewandte Chemie, International Edition. 2003;**42**:5548-5554. DOI: 10.1002/anie.200301652

[19] Bernal Ramos K, Saly MJ, Chabal YJ. Precursor design and reaction mechanisms for the atomic layer deposition of metal films. Coordination Chemistry Reviews. 2013;**257**:3271-3281. DOI: 10.1016/j.ccr.2013.03.028

[20] Suntola T. Atomic layer epitaxy. Materials Science Reports. 1989;**4**:261-312. DOI: 10.1016/S0920-2307(89)80006-4

[21] Li Z, Gordon RG. Thin, continuous, and conformal copper films by reduction of atomic layer deposited copper nitride. Chemical Vapor Deposition. 2006;**12**:435-441. DOI: 10.1002/cvde.200606485

[22] Donders ME, Arnoldbik WM, Knoops HCM, Kessels WMM, Notten PHL. Atomic layer deposition of $LiCoO_2$ thin-film electrodes for all-solid-state li-ion micro-batteries. Journal of the Electrochemical Society. 2013;**160**:A3066-A3071. DOI: 10.1149/2.011305jes

[23] Mane AU, Shivashankar SA. Atomic layer chemical vapour deposition of copper. Materials Science in Semiconductor Processing. 2004;**7**:343-347. DOI: 10.1016/j.mssp.2004.09.094

[24] Li Z, Rahtu A, Gordon RG. Atomic layer deposition of ultrathin copper metal films from a liquid copper(I) amidinate precursor. Journal of the Electrochemical Society. 2006;**153**:C787. DOI: 10.1149/1.2338632

[25] Moon DY, Han DS, Shin SY, Park JW, Kim BM, Kim JH. Effects of the substrate temperature on the Cu seed layer formed using atomic layer deposition. Thin Solid Films. 2011;**519**:3636-3640. DOI: 10.1016/j.tsf.2011.01.346

[26] Pierson HO. Handbook of Chemical Vapor Deposition: Principles, Technology and Applications; Materials Science and Process Technology. New York: William Andrew Inc; 1999. ISBN 9780815517436

[27] Choy KL. Chemical vapour deposition of coatings. Progress in Materials Science. 2003;**48**:57-170. DOI: 10.1016/S0079-6425(01)00009-3

[28] Thurier C, Doppelt P. Platinum OMCVD processes and precursor chemistry. Coordination Chemistry Reviews. 2008;**252**:155-169. DOI: 10.1016/j.ccr.2007.04.005

[29] Kafizas A, Carmalt CJ, Parkin IP. CVD and precursor chemistry of transition metal nitrides. Coordination Chemistry Reviews. 2008;**252**:155-169. DOI: 10.1016/j.ccr.2007.04.005.

[30] Kim H, Bhandari HB, Xu S, Gordon RG. Ultrathin cvd Cu seed layer formation using copper oxynitride deposition and room temperature remote hydrogen plasma reduction. Journal of the Electrochemical Society. 2008;**155**:H496. DOI: 10.1149/1.2912326

[31] Maruyama T, Ikuta Y. Copper thin films prepared by chemical vapour deposition from copper dipivalylmethanate. Journal of Materials Science. 1993;**28**:5540-5542. DOI: 10.1007/BF00367827

[32] Creighton JR, Ho P. Introduction to Chemical Vapor Deposition (CVD). In: Park JH, editor. Chemical vapour

deposition. Ohio: ASM International; 2001. pp. 1-10

[33] Bhandari HB, Yang J, Kim H, Lin Y, Gordon RG, Wang QM, et al. Chemical vapor deposition of cobalt nitride and its application as an adhesion-enhancing layer for advanced copper interconnects. ECS Journal of Solid State Science and Technology. 2012;**1**:N79-N84. DOI: 10.1149/2.005205jss

[34] Gaskell SJ. Electrospray: Principles and practice. Journal of Mass Spectrometry. 1997;**32**:677-688. DOI: 10.1002/(SICI)1096-9888(199707)32:7<677:AID-JMS536>3.0.CO;2-G

[35] Chen CH, Buysman AAJ, Kelder EM, et al. Fabrication of LiCoO2thin film cathodes for rechargeable lithium battery by electrostatic spray pyrolysis. Solid State Ionics. 1995;**80**:1-4

[36] Kessick R, Fenn J, Tepper G. The use of AC potentials in electrospraying and electrospinning processes. Polymer. 2004;**45**:2981-2984. DOI: 10.1016/j.polymer.2004.02.056

[37] Lincot D, Guillemoles JF, Taunier S, Guimard D, Sicx-Kurdi J, Chaumont A, et al. Chalcopyrite thin film solar cells by electrodeposition. Solar Energy. 2004;**77**:725-737. DOI: 10.1016/j.solener.2004.05.024

[38] Yadav AA, Chavan UJ. Electrochemical supercapacitive performance of spray deposited Co_3O_4 thin film nanostructures. Electrochimica Acta. 2017;**232**:370-376. DOI: 10.1016/j.electacta.2017.02.157

[39] Mallik A, Ray BC. Evolution of principle and practice of electrodeposited thin film: A review on effect of temperature and sonication. International Journal of Electrochemistry. 2011;**2011**:1-16. DOI: 10.4061/2011/568023

[40] Zhen Q. Synthesis and electrochemical properties of advanced nanostructured electrode for lithium secondary batteries starting from layered compounds[Internet]. Nagoya Institute of Technology; 2013. Available from: http://id.nii.ac.jp/1476/00003024/

[41] Morales J, Sánchez L, Bijani S, Martínez L, Gabás M, Ramos-Barrado JR. Electrodeposition of Cu[sub 2]O: An excellent method for obtaining films of controlled morphology and good performance in Li-ion batteries. Electrochemical and Solid-State Letters. 2005;**8**:A159. DOI: 10.1149/1.1854126

[42] Bacsa R, Ravindranathan P, Dougherty JP. Electrochemical-hydrothermal synthesis of barium titanate thin films on titanium substrates. Journal of Materials Research. 1996;**7**:423-428

[43] Davis JR, editor. Copper and Copper Alloys. Ohio, USA: ASM international; 2001. ISBN 0-87170-726-8

[44] Quan Z, Ohguchi S, Kawase M, Tanimura H, Sonoyama N. Preparation of nanocrystalline $LiMn_2O_4$ thin film by electrodeposition method and its electrochemical performance for lithium battery. Journal of Power Sources. 2013;**244**:375-381. DOI: 10.1016/j.jpowsour.2012.12.087

[45] Han S, Lee TL, Yang CJ, Shih HC. Trench gap-filling copper by ion beam sputter deposition. Materials Chemistry and Physics. 2006;**97**:19-22. DOI: 10.1016/j.matchemphys.2005.05.042

[46] Livage J. Sol-gel chemistry and electrochemical properties of vanadium oxide gels. Solid State Ionics. 1996;**86-88**:935-942. DOI: 10.1016/0167-2738(96)00336-0

[47] Dimitriev Y, Ivanova Y, Iordanova R. History of sol–gel science and technology. Journal of the University of Chemical Technology and Metallurgy. 2008;**34**:181-192

[48] Chen D. Anti-reflection (AR) coatings made by sol–gel processes: A review. Solar Energy Materials & Solar Cells. 2001;**68**:313-336. DOI: 10.1016/S0927-0248(00)00365-2

[49] Nagai H, Sato M. Heat treatment in molecular precursor method for fabricating metal oxide thin films. In: Czerwinski F, editor. Heat Treatment—Conventional and Novel Applications. Rijeka: InTech; 2012. pp. 297-322

[50] Nagai H, Aoyama S, Hara H, Mochizuki C, Takano I, Honda T, et al. Photoluminescence and photoreactivity affected by oxygen defects in crystal-oriented rutile thin film fabricated by molecular precursor method. Journal of Materials Science. 2010;**45**:5704-5710. DOI: 10.1007/s10853-010-4640-z

[51] Nagai H, Mochizuki C, Hara H, Takano I, Sato M. Enhanced UV-sensitivity of vis-responsive anatase thin films fabricated by using precursor solutions involving Ti complexes. Solar Energy Materials & Solar Cells. 2008;**92**:1136-1144. DOI: 10.1016/j.solmat.2008.04.005

[52] Likius DS, Nagai H, Aoyama S, Mochizuki C, Hara H, Baba N, et al. Percolation threshold for electrical resistivity of Ag-nanoparticle/titania composite thin films fabricated using molecular precursor method. Journal of Materials Science. 2012;**47**:3890-3899. DOI: 10.1007/s10853-011-6245-6

[53] Nagai H, Suzuki T, Takahashi Y, Sato M. Photovoltaic lithium-ion battery fabricated by molecular precursor method. Functional Materials Letters. 2016;**9**:1650046-1650041, 4. DOI: 10.1142/S1793604716500466

[54] Mochizuki C, Hara H, Takano I, Hayakawa T, Sato M. Application of carbonated apatite coating on a Ti substrate by aqueous spray method. Materials Science and Engineering: C. 2013;**33**:951-958. DOI: 10.1016/j.msec.2012.11.027

[55] Hishimone PN, Nagai H, Morita M, Sakamoto T, Sato M. Highly-conductive and well-adhered cu thin film fabricated on quartz glass by heat treatment of a precursor film obtained via spray-coating of an aqueous solution involving Cu(II) complexes. Coatings. 2018;**8**:352. DOI: 10.3390/coatings8100352

[56] Sato M, Hara H, Nishide T, et al. A water-resistant precursor in a wet process for TiO2 thin film formation. Journal of Materials Chemistry. 1996;**6**:1767

[57] Nagai H, Sato M. Highly functionalized lithium-ion battery. In: Yang D, editor. Alkali-ion Batteries. Rijeka: InTech; 2016. pp. 111-124

[58] Nagai H, Suzuki T, Mochizuki C, Takano I, Honda T, Sato M. Formation mechanism of *p*-type Cu_2O thin films via intermediate Cu^0 species derived from Cu(II) complex of ethylenediamine-*N*,*N*,*N′*,*N′*-tetraacetic acid. Science of Advanced Materials. 2014;**6**:603-611. DOI: 10.1166/sam.2014.1788

[59] Nagai H, Mita S, Takano I, Honda T, Sato M. Conductive and semi-transparent Cu thin film fabricated using molecular precursor solutions. Materials Letters. 2015;**141**:235-237. DOI: 10.1016/j.matlet.2014.11.056

[60] Nagai H, Suzuki T, Nakano T, Sato M. Embedding of copper into submicrometer trenches in a silicon substrate using the molecular precursor solutions with copper. Materials Letters. 2016;**182**:206-209. DOI: 10.1016/j.matlet.2016.06.123

[61] Khim D, Baeg K, Yu B, Kang S, Kang M. Spray-printed organic field-effect transistors and complementary inverters. Journal of Materials Chemistry C. 2013;**1**:1500-1506. DOI: 10.1039/c2tc00085g

[62] Eslamian M. Spray-on thin film pv solar cells: Advances, potentials and challenges. Coatings. 2014;**4**:60-84. DOI: 10.3390/coatings4010060

Section 3

Thin Film for Cathode

Chapter 3

Cathode Electronic Structure Impact on Lithium and Sodium Batteries Parameters

Janina Molenda

Abstract

The author of this work basing on her own investigations of A_xMO_2 cathode materials (A = Li, Na; M = 3d) has demonstrated that the electronic structure of these materials plays an important role in the electrochemical intercalation process. The proposed electronic model of intercalation is universal and has outstanding significance with regard to tailoring the properties of electrode materials to the most efficient application in Li-ion and Na-ion batteries. The paper reveals correlation between electronic structure, transport, and electrochemical properties of layered Li_xCoO_2, $Li_xNi_{1-y-z}Co_yMn_zO_2$ and Na_xCoO_2 cathode material and explains of apparently different character of the discharge/charge curve in Li_xCoO_2 (monotonous curve) and $NaxCoO_2$ systems (step-like curve). Comprehensive experimental studies of physicochemical properties of $Li_xNi_{1-y-z}Co_yMn_zO_2$ cathode material (XRD, electrical conductivity, and thermoelectric power) are supported by electronic structure calculations performed using the Korringa-Kohn-Rostoker method with the coherent potential approximation (KKR-CPA) to account for chemical disorder. It is found that even small oxygen defects (~1%) may significantly modify DOS characteristics via formation of extra broad peaks inside the former gap leading to its substantial reduction.

Keywords: lithium and sodium intercalation, electronic structure, Li_xCoO_2, $Li_xNi_{1-y-z}Co_yMn_zO_2$, Na_xCoO_2

1. Introduction

Lithium or sodium intercalation into layered MO_2 transition metal oxides involves both ions and electrons, and can be expressed by the equations:

$$xLi^+ + xe^- + MO_2 \rightleftarrows Li_xMO_2 \quad \text{or} \quad xNa^+ + xe^- + MO_2 \rightleftarrows Na_xMO_2 \tag{1}$$

This reaction involves energy at the level of several eV/atom, which is associated with deep *d*-type electron levels in transition metal compounds. This enables the storage of several hundred Wh/kg and makes it possible to build power sources with high volumetric and gravimetric energy density.

The many years of the author's studies of various transition metal compounds: Li_xTiS_2 and Na_xTiS_2 [1], Li_xCoO_2 [2], Na_xCoO_2 [3–6], Li_xVO_2 [7, 8], $Li_xW_yV_{1-y}O_2$ [9], $LiNiO_2$ [10], Li_xWO_3 [11], $Li_xYBa_2Cu_3O_{7-\delta}$ [12, 13], $Li_xMn_2O_4$ [14, 15], $Li_x(Co,Ni,Mn)O_2$ [16], $LiNi_{0.5-y}Cu_yMn_{1.5}O_{4-\delta}$ [17] showed that the electronic structure and transport properties of the cathode material play a significant role in the intercalation process. Experience in the field of ionic and electronic defect structure in transition metal compounds allowed the author to see the phenomenon of intercalation as an ionic-electronic process from a perspective different than the one that is commonly presented in the literature. This became a starting point for the proposition of an original concept of the intercalation process and the related phenomena [3, 14, 18, 19].

For a $A/A^+/A_xMO_2$-type cell (A = Li or Na) with a cathode material with the concentration of electronic charge carriers in A_xMO_2 determined by the concentration of intercalated alkaline ions, it can be demonstrated that the variation of the cell's electromotive force as a function of intercalation degree corresponds to variations of the Fermi level in the cathode material [4, 14]. The electromotive force (E) of a $A/A^+/A_xMO_2$ cell is the difference in the chemical potential of lithium (sodium) in the cathode and anode (metallic Li or Na) materials:

$$\mu_{A(cathode)} - \mu_{A(anode)} = -F \cdot E \tag{2}$$

Since the potential of the A/A^+ anode is constant (constant concentration of Li^+ or Na^+ ions in electrolyte), the variations of electromotive force of the cell can be ascribed to the changes in $\mu_{A(cathode)}$, i.e. $-F \cdot \Delta E = \Delta\mu_{A(cathode)}$. The chemical potential of lithium (sodium) in the cathode material can be represented as a sum of chemical potentials of lithium (sodium) ions and electrons:

$$\mu_{A(cathode)} = \mu_{A^+} + \mu_e \tag{3}$$

Since the chemical potential of lithium (sodium) ions can be expressed as:

$$\mu_{A^+} = \mu^o{}_{A^+} + k_B T \cdot \ln[A^+] \tag{4}$$

thus:

$$\Delta\mu_{A^+} = k_B T \ln\frac{[A^+]_f}{[A^+]_i} \tag{5}$$

where *f* and *i* denote the final and initial states, respectively.

The chemical potential of electrons in the cathode material can be identified as the energy of the Fermi level; the changes in the latter are determined by the electronic structure (DOS—density of states) in the vicinity of E_F, and can vary in the range of 1 eV or more upon the introduction of electrons during lithium (sodium) intercalation. For a change in lithium (sodium) ion concentration that is of the order of 1 mole ($\Delta[A^+] \sim 1$), the change in the chemical potential of lithium (sodium) ions in the cathode material ($\Delta\mu_{A^+}$) is of the order of k_BT, i.e. only 0.025 eV at the room temperature (Eq. (5)), while the change in the chemical potential of electrons in the cathode material may be as much as two orders of magnitude higher ($\Delta\mu_{A^+} << \Delta\mu_e$). Therefore, the variations of the electromotive force of the $A/A^+/A_xMO_2$ cell which accompany the intercalation reaction correspond mainly to those in the chemical potential of electrons (i.e. Fermi level variations) of the cathode material. **Figure 1** shows the electronic scheme of the $A/A^+/A_xMO_2$ cell and depicts the difference in the chemical potentials of electrons

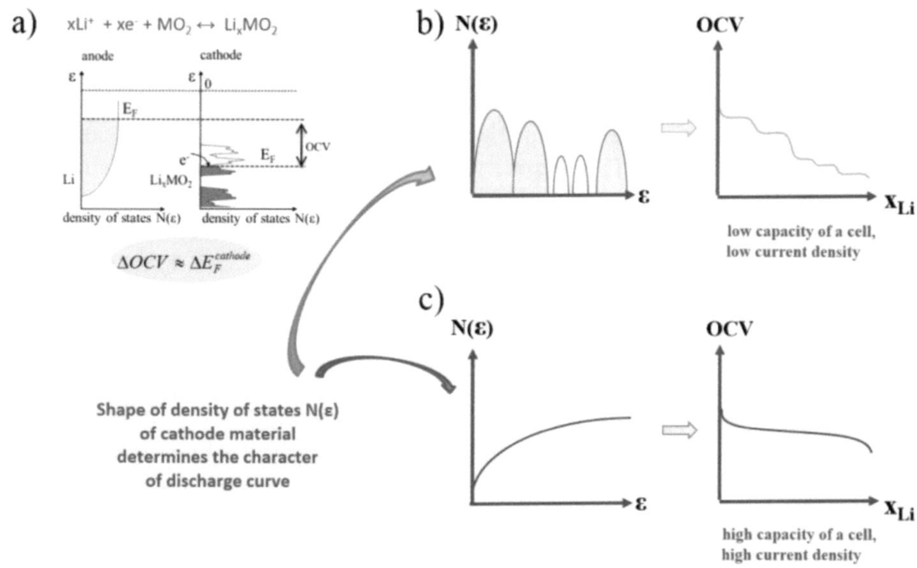

Figure 1.
(a) Density of states of Li_xMO_2 and lithium illustrating difference in chemical potential of electrons and resulting electromotive force of $Li/Li^+/Li_xMO_2$ cell. Expected character of the discharge curve (EMF) of the $Li/Li^+/Li_xMO_2$ cell, depending on the electronic structure of the cathode material (step-like (b) and monotonic (c)).

in the cathode and anode materials and the related electromotive force of the cell. The electrons inserted into the cathode material during intercalation together with an equivalent number of lithium (sodium) ions (Eq. (1)) occupy the available electronic states at the Fermi level and raise it in a way dependent on the profile of the density of states function.

The character of the density of states function determines the shape of the discharge curve (monotonic or step-like). **Figure 1b** and **c** illustrate the correlation between the electronic structure of a cathode material and the nature of its discharge curve. A continuous density of states function in a cathode material leads to a monotonic discharge curve that is beneficial from the point of view of practical application (**Figure 1c**), while a discontinuous density of states function leads to an adverse, step-like nature of the discharge curve (**Figure 1b**). In addition, the localization of electronic states that accompany the discontinuous density of states function limits the kinetics of the lithium intercalation process, reducing the current density of a cell and making the utilization of the theoretical capacity of a cathode material impossible.

The proposed electronic model of electrochemical intercalation explains both the monotonic and step-like characters of the discharge curve, and allows the anticipation and engineering of the properties of intercalated electrode materials. The presented model also demonstrates that a method of studying the Li^+/Li_xMO_2 cathode potential by measuring the electromotive force of a $Li/Li^+/Li_xMO_2$ cell is an excellent tool with regard to experimental solid state physics, allowing the direct observation of changes in the position of the Fermi level in Li_xMO_2 during "lithium intercalation." Similar conclusions are valuable for Na_xMO_2 systems.

Below we present three examples of intercalated transition metal oxides Li_xCoO_2 [5, 20], $Li_x(Co,Ni,Mn)O_2$ [21] and Na_xCoO_2 [3–5], in which through the intercalation of alkali metal the controlled crossing insulator-metal can be performed, and track changes in the position of the Fermi level as a function of concentration of electrons introduced in the intercalation process (along with alkali ions).

2. Li_xCoO_2

$LiCoO_2$ is currently used as a cathode material in commercial Li-ion batteries, due to, among other advantages, its high voltage. Cycling with upper cut off set at 4.2 V corresponds to deintercalation/intercalation of about 0.5 Li per $LiCoO_2$ formula unit, and gives a specific capacity of about 140 mAh g^{-1}. Deeper deintercalation causes structural instability of the Li_xCoO_2 cathode material, loss of oxygen from the material, and its reactivity with liquid electrolyte [22–25].

$LiCoO_2$ exhibits hexagonal-type lattice (O3 structure) with R-3 m space group symmetry (**Figure 2**). This structure consists of close-packed oxygen layers stacked in ABC-like sequence. Cobalt and lithium ions are surrounded by six oxygen atoms, forming CoO_6 and LiO_6 octahedral coordination. CoO_6 octahedra share edges creating two-dimensional arrangements (CoO_2 layers), between which van der Waals-type gap exists, occupied by lithium ions. Extraction of lithium from $LiCoO_2$ proceeds as a sequence of the phase transitions. For lithium content $0.95 \leq x \leq 1$ one-phase solid solution domain with hexagonal structure (hex-I) is observed. Two-phase coexistence region is found to exist between Li concentrations of x = 0.94 and 0.75 [20, 26–30]. These two phases share the same crystal symmetry, but show difference in the lattice parameters, mainly in the c parameter. Many authors report that for lithium content below x = 0.75, one phase region with hex-II phase exists [26–30]. For x = 0.5, a transition from hexagonal to monoclinic phase occurs, which is accompanied by a large increase of the unit cell dimension along the c-axis, up to 1.7%.

Our results, presented in papers [20], diverge with the mentioned literature data, where it is stated that hex-II phase replaces hex-I phase for the compositions with x < 0.75. We also do not agree with a thesis that hex-I phase exhibits semi-conducting properties, while hex-II phase shows metallic-type behavior [20]. It is known that upon lithium deintercalation crystal structure of $LiCoO_2$ undergoes

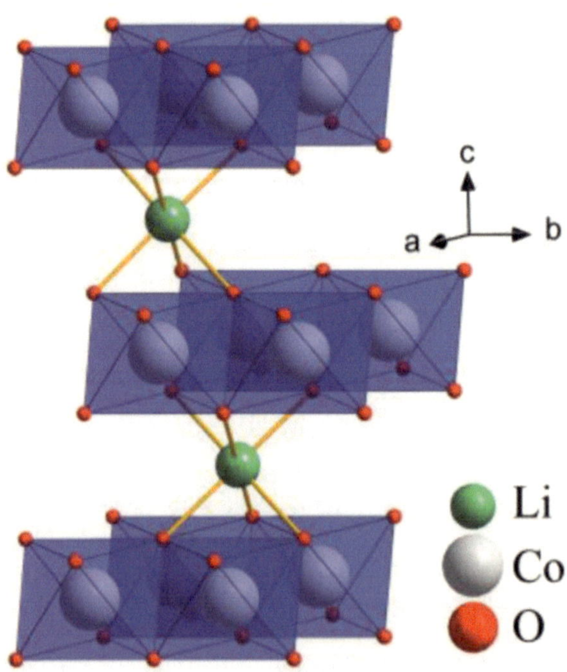

Figure 2.
Hexagonal O3 structure of $LiCoO_2$ with the R-3m space group.

modification. **Figure 3a** presents variations of lattice parameters of hexagonal phase of Li_xCoO_2 during deintercalation of Li. **Figure 3b** presents the mole fraction of the hex-II phase as a function of lithium content.

One can see that hex-II phase appears for $x = 0.94$ but fast disappears on further deintercalation. **Figure 3c** presents z parameter of oxygen 6*c* (0,0,z) position, responsible for a deformation of oxygen octahedron surrounding Co cation. While lattice parameters a and c evolve on deintercalation only slightly, z parameter changes in irregular way. Parameter z and its importance on the electronic structure of Li_xCoO_2 will be discuss later. In Li_1CoO_2 Co^{3+} cations ($3d^6$) adopt low-spin configuration ($t_{2g}^6e_g^0$), being present in octahedral oxygen environment. Cobalt 3*d* orbitals are energetically close to oxygen 2*p* orbitals, and as a result, considerable covalent mixing, and consequently, a high degree of delocalization of the electronic states over the CoO_6 octahedra occurs [31]. Deintercalation of lithium ions from $LiCoO_2$ structure is accompanied by charge compensation at the Co sites, and leads to a formation of Co^{4+} ($3d^5$) states. Above 4.2 V, energy level of $Co^{3+/4+}$ redox reaction overlaps with the top of the O2p band, and as a results, to the oxidation and release of the oxygen from the structure of the cathode material [31].

In this work, electronic structure calculations for Li_xCoO_2 system were performed using the Green function Korringa-Kohn-Rostoker method with the coherent potential

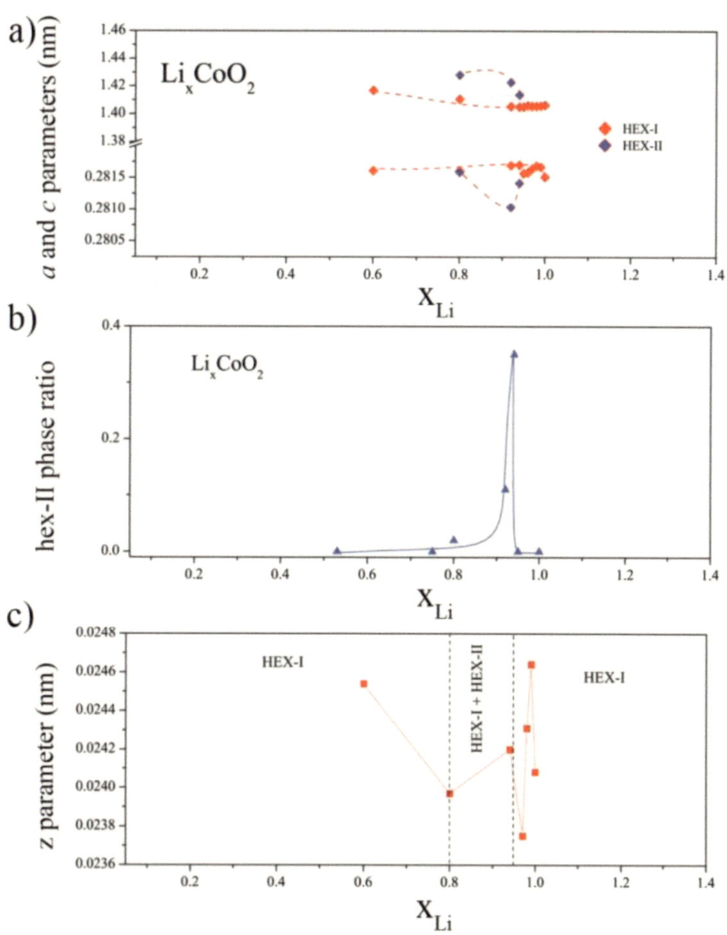

Figure 3.
(a) Variations of lattice parameters of hexagonal Li_xCoO_2 phase during deintercalation of lithium, (b) the mole fraction of hex-II phase and (c) changes of z parameter during deintercalation of lithium.

approximation (KKR-CPA), which allows to account for chemical disorder [20, 32–34]. **Figure 4** presents such calculations for the stoichiometric $LiCoO_2$ and for the deintercalated Li_xCoO_2 with x = 0.99, 0.97 and 0.6. Electronic structure of the starting $LiCoO_2$ (**Figure 4a**) consists of valence and conductivity band, separated by an energy gap of the order of 1 eV. The valence states are formed essentially from strongly hybridized *d*-states of Co and *p*-states of oxygen. Let us consider the following:

What are electronic structure predictions about the properties of Li_xCoO_2 system during deintercalation of lithium i.e. during extraction of electrons?

What about evolution of the Fermi level position and its influence on a character of the OCV curve changes occurring during charge of $Li/Li^+/Li_xCoO_2$ cell?

What about modification of transport properties of Li_xCoO_2 in this process?

For starting, stoichiometric $LiCoO_2$, the Fermi level is situated in the energy gap (**Figure 4a**), so semiconducting properties should be observed. During deintercalation of lithium, the electrons are extracted from the valence band, and the Fermi level is sharply shifted to a new position in the valence band **Figure 4b**), while further changes to its placement, occurring during the deintercalation process should be monotonous in a wide range of lithium concentration (**Figure 4b–d**). Therefore, electrical properties of Li_xCoO_2 should become more and more metallic, due to the shifting of the Fermi level in these regions, where sharp increase of the density of states appears.

In order to verify our electronic structure predictions on transport and electrochemical properties of Li_xCoO_2 system, comprehensive studies of the OCV charge curve and transport properties were conducted. **Figure 5** present charge curve (OCV) for Li_xCoO_2. Points denotes lithium concentrations, for which the work of the

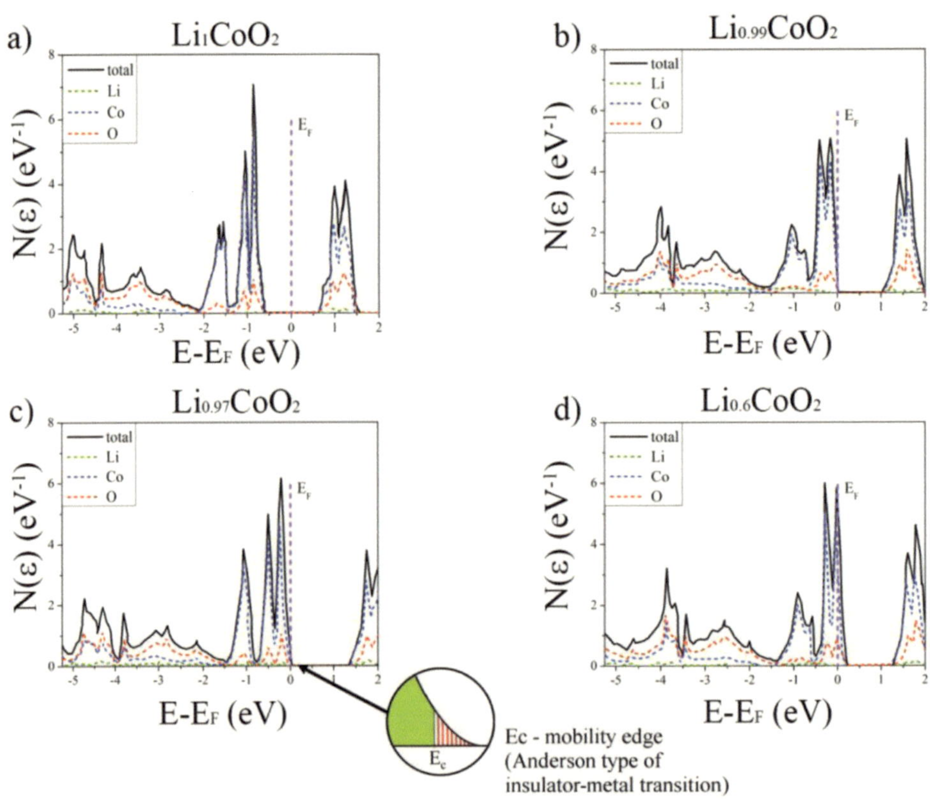

Figure 4.
Electronic structure calculations (KKR-CPA) for starting $LiCoO_2$ and deintercalated Li_xCoO_2 for (a) x = 1, (b) x = 0.99, (c) x = 0.97 and (d) x = 0.6.

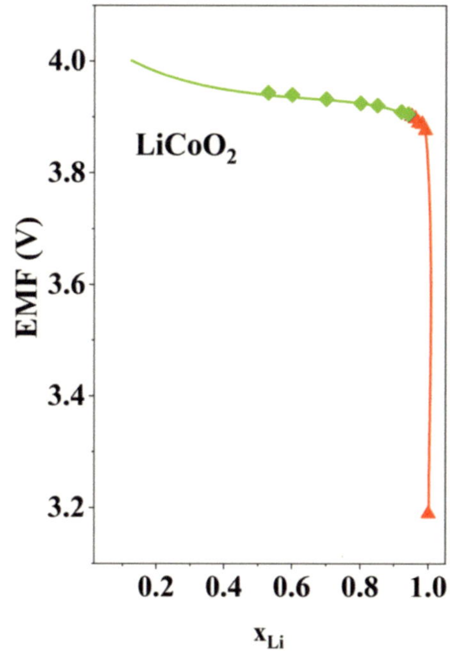

Figure 5.
Li_xCoO_2 charge curve (OCV).

cathode material was stopped in order to examine its structural properties, conduct NEXAFs measurements, as well as perform electrical conductivity and thermoelectric power studies. **Figure 6a** and **b** present temperature dependence of the electrical conductivity and thermoelectric power for Li_xCoO_2 system. Comparison and analysis of **Figures 5–7** show strong correlation between electronic structure, Fermi

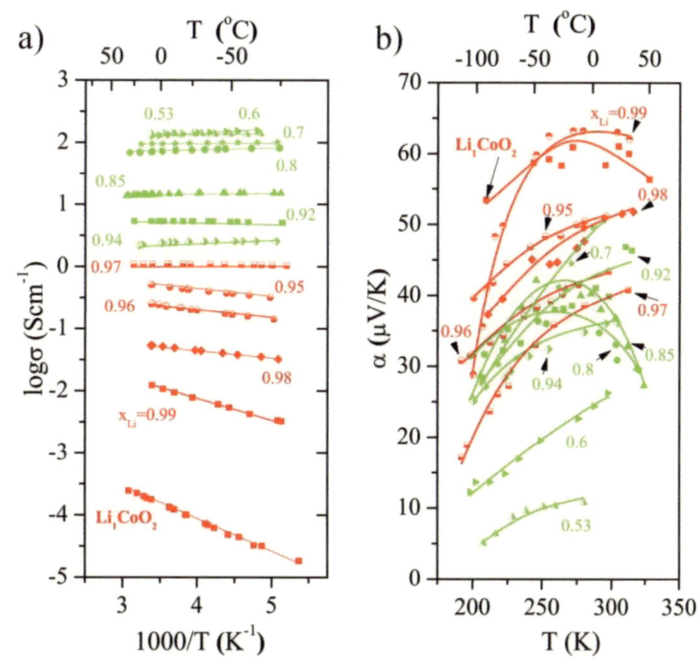

Figure 6.
Temperature dependence of (a) electrical conductivity and (b) thermoelectric power for Li_xCoO_2 system.

level variations, character of the charge/discharge curve and transport properties of Li_xCoO_2 system. For starting material, Li_1CoO_2, Fermi level is situated in the energy gap (**Figure 4a**), so we observed semiconducting-like properties (**Figure 6a**), and a potential jump occurring at the beginning of the OCV charge curve (**Figure 5**).

Deintercalation of lithium (related to an extraction of electrons from the valence band) leads to a fast movement of the Fermi level to the valence band, followed by a monotonous changes in a wide range of lithium concentration (monotonous part of the charge curve **Figure 5**). According to our predictions, electrical properties modify toward metallic-like ones. Only for $x_{Li} = 1$ the activation energy of the electrical conductivity can be related directly to the energy gap, while for $0.94 < x_{Li} < 1$, the activation energy of the electrical conductivity is related to the activation energy of mobility of electron holes in the region of the tail of the valence band. Therefore, we postulate Anderson-type insulator-metal transition, related to a chemical disorder present in this system. The states above mobility edge E_c are localized, while these below E_c are delocalized (see insert in **Figure 4c**). For $x = 0.94$ Fermi level crosses the mobility edge E_c.

Analysis of the electronic structure of Li_xCoO_2 leads to the ascertainment that valence states do not evolve in a regular way with Li concentration (**Figure 4a–d**). Electronic spectrum for $x = 0.97$ seems to be more similar to that one for $x = 1$. To have a deeper insight into possible reasons of such peculiar behavior of the electronic structure of Li_xCoO_2 we analyzed crystallographic parameters namely, parameter z indicating position of the oxygen anions in the unit cell. It was found that the z parameter changes irregularly with lithium concentration (**Figure 3c**). We computed DOS for $LiCoO_2$ for different z using experimental values taken from the crystallographic data, as shown in **Figure 7** [20]. One can see that z parameter affect strongly valence DOS shape of $LiCoO_2$ oxide.

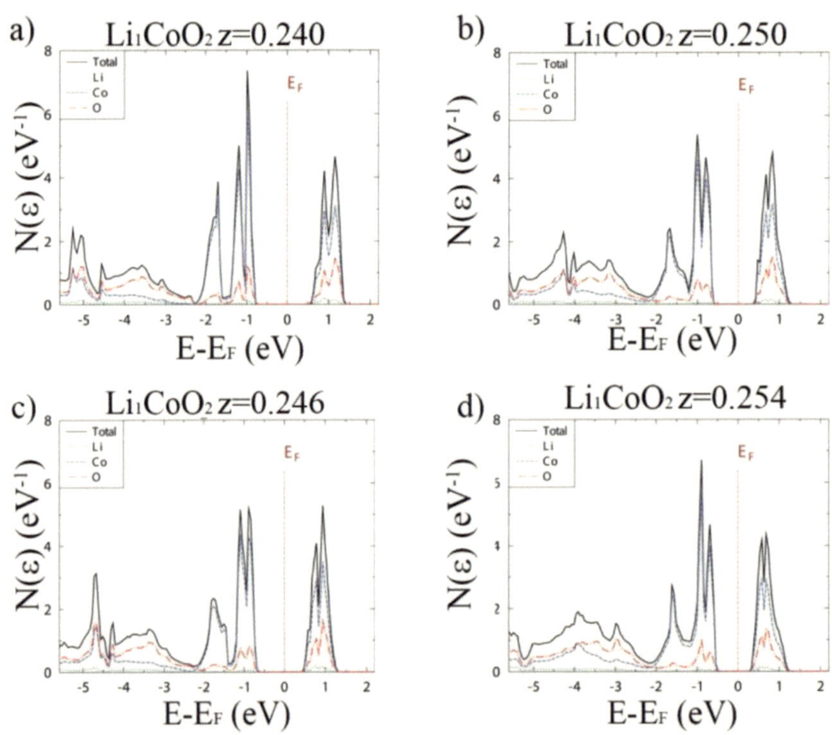

Figure 7.
(a–d) Electronic structure calculations (KKR-CPA) for $LiCoO_2$ for different z parameters (a) $z = 0.240$, (b) $z = 0.246$, (c) $z = 0.250$, and (d) $z = 0.254$.

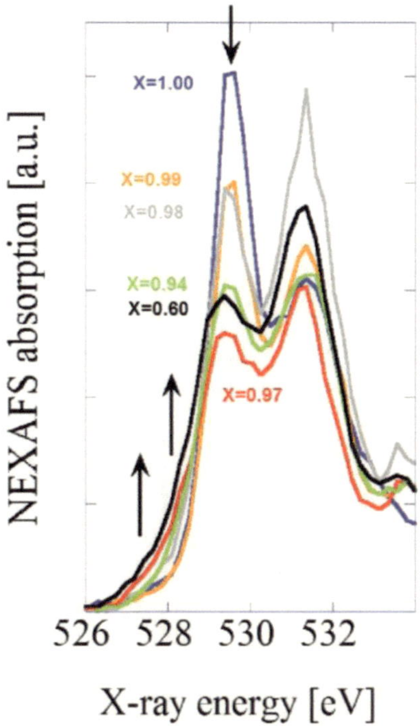

Figure 8.
NEXAFS oxygen 1 s spectra for Li_xCoO_2 samples with lithium concentration x = 1.00, 0.99, 0.98, 0.97, 0.94 and 0.60.

Figure 8 shows NEXAFS oxygen 1 s spectra of the Li_xCoO_2 samples with lithium concentration x = 1.00, 0.99, 0.98, 0.97, 0.94 and 0.60 [20]. The NEXAFS measurements of Li_xCoO_2 show a radical evolution of the top valence band features with the lithium deintercalation degree. These results corroborate well with crystallographic data, as well as with the particular effect of the oxygen octahedron distortion on the computed DOS shape (note that the respective peaks are presented in a different order on energy scale in **Figures** 7 and **8**). We suppose that the z parameter (determining position of O sites), which is responsible for the deformation of oxygen octahedron surrounding Co atoms, is also responsible for modification of valence DOS, and is a reason of major discrepancies in the observed crystallographic parameters and phase diagram of Li_xCoO_2 system, presented by different authors.

3. $LiNi_{1-y-z}Co_yMn_zO_2$

To correlate electrochemical properties of $Li_xNi0_{.65}Co_{0.25}Mn_{0.1}O_2$ and $Li_xNi_{0.55}Co_{0.35}Mn_{0.1}O_2$ mixed cathode materials with their transport and electronic structure properties, we stopped the work of the cathode material at the characteristic points of the charge curve (EMF, **Figure 9a**) in order to examine its properties as a function of lithium content.

The evolution of a and c parameters during the lithium deintercalation of $Li_xNi_{0.65}Co_{0.25}Mn_{0.1}O_2$ and $Li_xNi_{0.55}Co_{0.35}Mn_{0.1}O_2$ oxides are presented in **Figure 9b**. The a parameter decreases due to oxidation of the transition metal ions (Ni and Co) and c parameter increases due to the increase in electrostatic repulsion between $((Ni,Co,Mn)O_2)n$ sheets. The monotonic changes of the lattice parameters indicate that the crystal structure is maintained during electrochemical deintercalation of

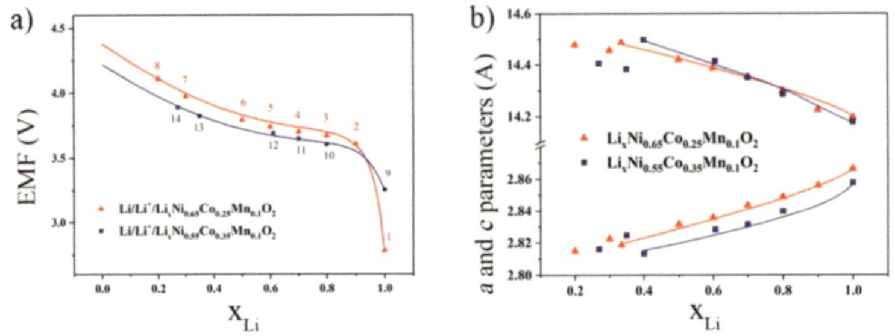

Figure 9.
(a) EMF curves for Li/Li⁺/LixNio.65Coo.25Mno.1O2 and Li/Li⁺/LixNio.55-Coo.35Mno.1O2 cells. Points 1–14 denote lithium concentration, for which the work of the cathode material was stopped in order to study its physicochemical properties. (b) Unit cell parameters vs. Li content in $Li_xNi_{0.65}Co_{0.25}Mn_{0.1}O_2$ and $Li_xNi_{0.55}Co_{0.35}Mn_{0.1}O_2$ cathode materials.

lithium. However for lithium content lower than 0.4 mol mol^{-1}, the c parameter starts to decrease and the a parameter starts to increase that is caused due to the structural instability of the deintercalated $Li_xNi_{0.9-y}Co_yMn_{0.1}O_2$ materials in this range (oxygen ions oxidize, leave two electrons and molecular oxygen evolves from the crystal structure). Higher structural stability is observed for oxides with higher nickel content.

The temperature dependences of electrical conductivity for electrochemically deintercalated $Li_xNi_{0.65}Co_{0.25}Mn_{0.1}O_2$ and $Li_xNi_{0.55}Co_{0.35}Mn_{0.1}O_2$ oxides are presented in **Figure 10a** and **b**. The collected data show that electrical conductivity decreases with decreasing lithium concentration, and it retains its thermally activated character with activation energy of the order of 0.15–0.25 eV in the entire lithium deintercalation range. However for lithium content near x_{Li} = 0.3, the conductivity starts to increase (**Figure 10a** and **b**). **Figure 11** presents the evolution of DOS for $Li_xNi_{0.65}Co_{0.25}Mn_{0.1}O_{2-y}$ from KKR-CPA calculations for selected Li content in the presence of oxygen vacancy as low as y = 0.01 [21]. From precise oxygen nonstoichiometry measurements performed for $Li_xNi_{0.65}Co_{0.25}Mn_{0.1}O_{2-y}$ oxide synthesized in air at the temperature of 850°C and quenched to room temperature, the oxygen nonstoichiometric value (y) could be estimated to be 0.06.

Our thermogravimetric studies show a much lower value 0.014. It is found that such defects strongly affect the electronic states particularly in the vicinity of the conductivity band edge, forming three additional DOS peaks inside the gap. Due to high degree of chemical disorder on transition metal sites, the observed 'defect' states are quite broad on comparing with the corresponding results in Li_xCoO_2 [20]. Interestingly, the KKR-CPA calculations deliver two important results: (i) the 'defect'

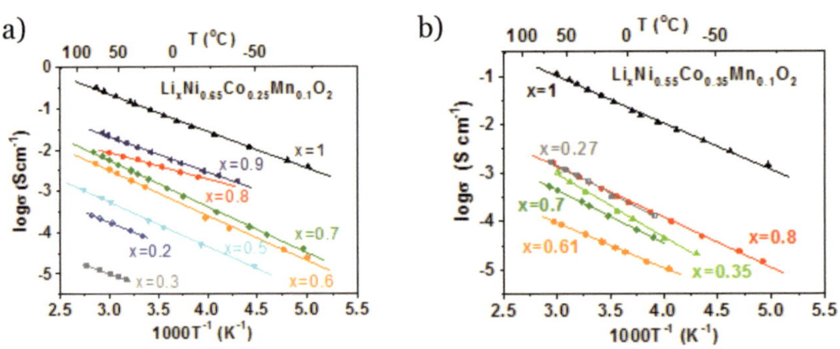

Figure 10.
Temperature dependence of the electrical conductivity of partially deintercalated (a) $Li_xNi_{0.65}Co_{0.25}Mn_{0.1}O_2$, (b) $Li_xNi_{0.55}Co_{0.35}Mn_{0.1}O_2$ oxides.

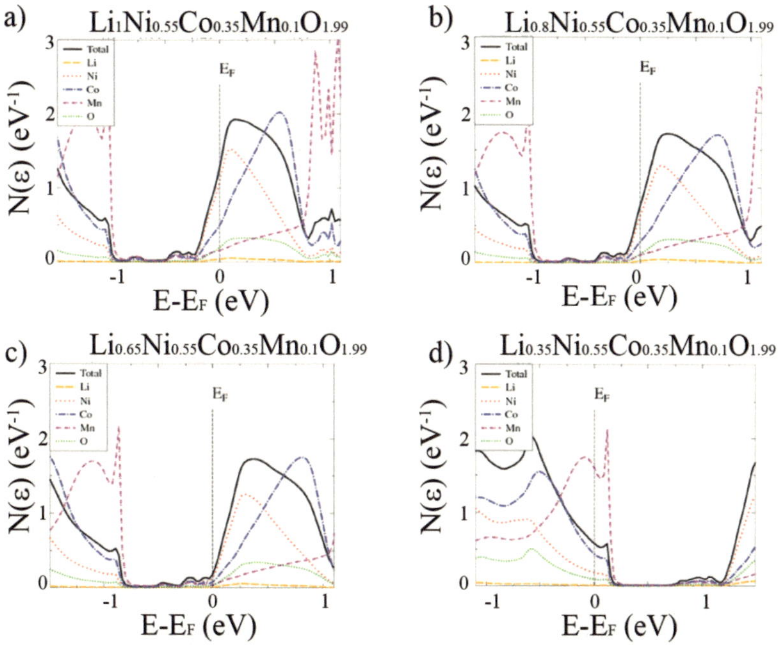

Figure 11.
(a–d) DOS in $Li_xNi_{0.65}Co_{0.25}Mn_{0.1}O_{2-y}$ for x = 1.0, 0.8, 0.65 and 0.35 when accounting for oxygen vacancy concentration y = 0.01, as calculated by the KKR-CPA method. Total and site-decomposed DOS (per atom) are plotted (see legend). The Fermi energy (E_F) is shifted to zero. Extra peaks inside the energy are clearly observed.

states mostly contain d-states of Ni, Co and Mn, not p-states of O and, (ii) the DOS details evolve remarkably with varying Li content. Hence, it is difficult to estimate whether, in such a disordered system, the Fermi level could fall into the real gap at any Li concentration, but surely the oxygen vacancies substantially reduce the energy gap seen on DOS of $LiCoO_2$ [5, 20] and $Li_xNi_{0.9-y}Co_yMn_{0.1}O_2$ (**Figure 11**). Thus, the activation character of measured electrical conductivity could be rather related to the pseudo-gap behavior observed on DOS.

When plotting DOS at the Fermi level in the two aforementioned series of oxides (**Figure 12a**) with varied Li concentrations, we observe the finite values of DOS for all considered compositions with a deep minimum obtained for about 0.5–0.6 Li content. It is worth considering that DOS characteristics should have an important influence on electron transport properties, which are mainly governed by electronic states lying in close vicinity to the Fermi level. Indeed, the measured temperature-dependent electrical conductivity in $Li_xNi_{0.55}Co_{0.35}Mn_{0.1}O_2$ and $Li_xNi_{0.65}Co_{0.25}Mn_{0.1}O_2$

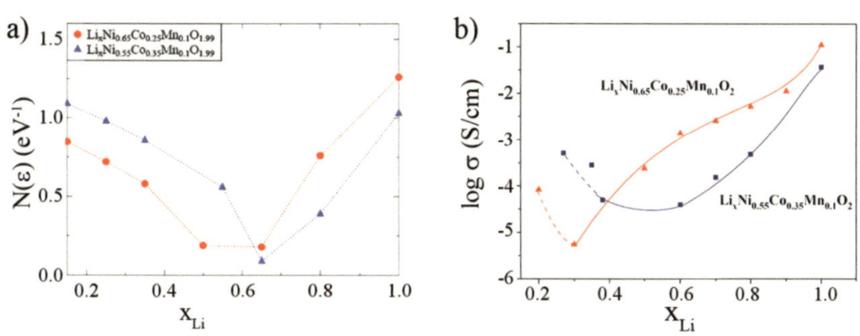

Figure 12.
(a) Variation of DOS at EF in $Li_xNi_{0.65}Co_{0.25}Mn_{0.1}O_{2-y}$ and $Li_xNi_{0.55}Co_{0.35}Mn_{0.1}O_{2-y}$ and (b) conductivity with varying Li content, when accounting for oxygen vacancy concentration y = 0.01.

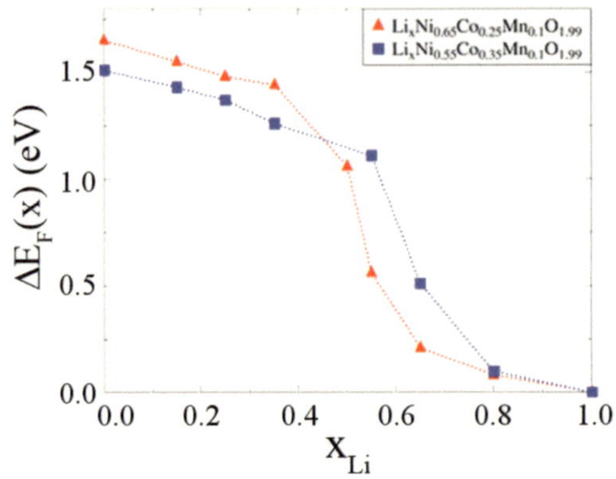

Figure 13.
Relative variation of Fermi energy (with respect to x = 1) with Li content of $Li_xNi_{0.55}Co_{0.35}Mn_{0.1}O_{2-y}$ and $Li_xNi_{0.65}Co_{0.25}Mn_{0.1}O_{2-y}$ cathode materials.

exhibit rather irregular behaviors with varying Li content (**Figure 10**), which could be tentatively interpreted by the variable character of the corresponding DOS curves. Such electronic structural behaviors cause both types of charge carriers, electrons and holes, to conduct electrical current in these materials. Furthermore, the character of electrical conductivity curves is mainly guided by competition between semiconducting like vs. metallic-like character.

Figure 12b shows variation of electrical conductivity of $Li_xNi_{0.65}Co_{0.25}Mn_{0.1}O_{2-y}$ and $Li_xNi_{0.55}Co_{0.35}Mn_{0.1}O_{2-y}$ in function of lithium content at room temperature. Strong correlation between the electronic structures of layered $Li_xNi_{0.9-y}Co_yMn_{0.1}O_{2-y}$ cathode materials and its electron transport properties was evidenced with characteristic minimum for $0.5 < x_{Li} < 0.6$ for both DOS and electrical conductivity (**Figure 12**).

The observed variations of the electromotive force of the $Li/Li^+/Li_xNi_{0.9-y}Co_yMn_{0.1}O_{2-y}$ cells are in good agreement with the calculated variations of chemical potential of electrons (i.e. the Fermi level variations) of the $Li_xNi_{0.9-y}Co_yMn_{0.1}O_{2-y}$ cathode material. **Figure 13** presents calculations of relative variation of Fermi energy (with respect to x = 1) with Li content. It is interesting to note that difference of the Fermi energy for the whole range of Li content is at the order of 1.5 eV, which corresponds well to the variation of EMF of the $Li/Li^+/Li_xNi_{0.9-y}Co_yMn_{0.1}O_{2-y}$ cell (**Figure 9a**).

4. Na_xCoO_{2-y}

Sodium analogue to $LiCoO_2$ has been extensively investigated as a potential candidate Na-intercalation-type cathode material [6, 35]. Na_xCoO_2 exhibits different electrochemical behavior, comparing to $LiCoO_2$ and shows subtle, but important differences in the crystal structure. Na_xCoO_2 presents step-like character of the discharge curve, which in according to electronic model of intercalation is surprising for the metallic system.

P2-type structure of Na_xCoO_2 ($x \approx 0.7$) consists of Na layers, which are stacked between layers of edge sharing CoO_6 octahedra (**Figure 14**) In this structure sodium occupies two different types of trigonal prismatic site: Na(1), which shares only faces with two CoO_6 octahedra of adjacent slabs, whereas Na(2) shares edges with the six surrounding CoO_6 octahedra. In addition, it can be expected that Na(1) site is less favorable energetically, due to a stronger electrostatic repulsion from the Co ions, but simultaneous occupation of Na(1) and Na(2) sites results in a

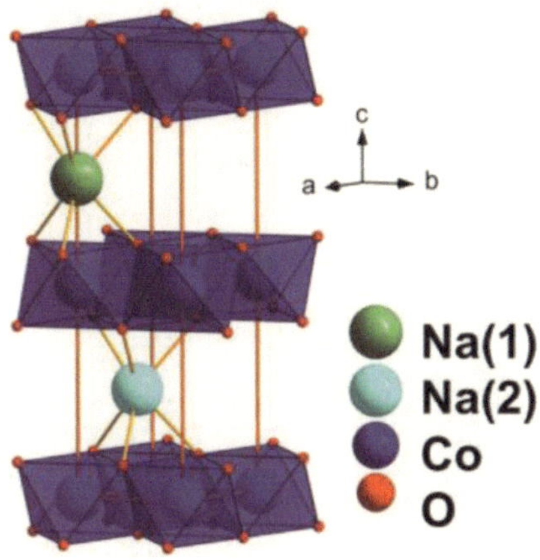

Figure 14.
Two different crystallographic sites of sodium in the Na_xCoO_{2-y} structure.

larger average Na-Na distance, comparing to a situation when only Na(2) sites are occupied. Changes of Na(1) to Na(2) ratio as a function of Na content give rise to a multitude of different types of Na ordering motifs, which can be grouped in three main pattern types: row-like, zigzag-like and droplet-like [36, 37].

Na_xCoO_2 is a potential material for thermoelectric applications due to high absolute values of thermoelectric power and metallic-type character of the electrical conductivity [38]. Sodium cobaltate exhibits superconductivity only in a narrow range of sodium content ($1/4 < x < 1/3$) below 4.5 K in the hydrated compound [39]. Occurrence of superconductivity is likely related to the lattice dimensionality and modification of the density of states near Fermi level, however, it seems that water molecules are not only passive lattice spacers [40].

The structural properties of the CoO_2 layers in Na_xCoO_2 and Li_xCoO_2 significantly affect electronic structure of these compounds. Generally, it can be stated that two-dimensional metallic CoO_2 layers, in which cobalt exists as Co^{3+} and Co^{4+} ions, which are separated by intercalating Li^+ or Na^+ ions layers, are a source of specific electronic, magnetic and electrochemical properties. Near $x = 0.5$, Co—O bond length and in-plane Co—Co distance are quite similar in Na_xCoO_2 and Li_xCoO_2 [41]. These parameters and thickness of CoO_2 layers increase significantly with the increasing x in Na_xCoO_2, while in Li_xCoO_2 they change only slightly.

It may be a consequence of a relatively large size of sodium cations (0.104 nm), which gives rise to a dramatic expansion of thickness of CoO_2 layers in Na_xCoO_2. According to the differences between geometry of CoO_2 layers in sodium and lithium system for $0.5 < x < 1.0$ range, it may be assumed that more intensive changes in the electronic band occupancy would occur in Na_xCoO_2 than in Li_xCoO_2. **Figure 15** presents step-like character of the OCV discharge curve for $Na/Na^+/Na_xCoO_{2-y}$ cell. The work of the cathode material was stopped in the characteristic points of the discharge curve, i.e. on the potential pseudo-plateau and on the potential jumps, in order to examine its structural, transport and thermal properties. Precise structural investigations of the Na_xCoO_{2-y} cathode material in equilibrium (fully relaxed cathode material) at characteristic points of a $Na/Na^+/Na_xCoO_{2-y}$ cell's discharge curve, i.e. on "potential pseudo-plateaus"[1] and potential jumps, pointed to a single-phase system in the sodium content range of 0.55–0.9, i.e. in the region where the step-like character of discharge curve is observed. The obtained results show no significant dependence of parameters a and c on the amount of sodium in the range where the

[1] Precise measurements of potential at the plateau indicate its decrease with sodium content.

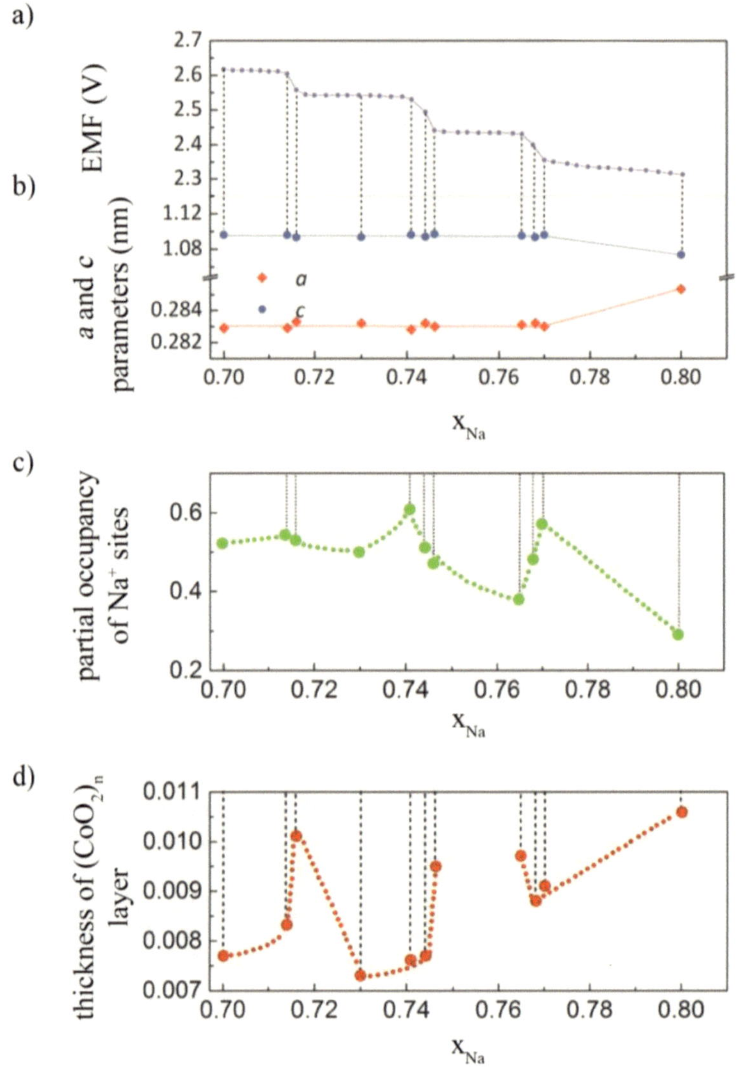

Figure 15.
(a) Step-like character of an OCV discharge curve for the $Na/Na^+/Na_{0.7}CoO_{2-y}$ cell, (b) evolution of lattice parameters with sodium content, (c) partial occupancy of two Na^+ sites, (d) parameter z indicating the distortion of CoO_6—oxygen octahedra.

potential jumps appear (**Figure 15b**); however, the occupancy of the two different Na sites and parameter *z* (thickness of CoO_2 layers, indicating CoO_6-octahedra distortion) exhibit maxima in the area of potential jumps (**Figure 15c** and **d**, respectively).

Below we present the results of electronic structure calculations of Na_xCoO_{2-y} performed by KKR-CPA technique (the Korringa-Kohn-Rostoker method combined with coherent potential approximation) [3, 32, 33], which take into account the chemical disorder in the system arising from the presence of oxygen vacancies and the two different crystallographic sites of sodium. Our calculations [3] show that there is no effect of sodium sublattice ordering on the electronic structure of $Na_{0.75}CoO_2$. For four different crystallographic configurations of sodium, only very small changes in the shape of the valence band are observed, and no extra defect bands in the energy gap. **Figure 16b** presents the influence of the non-stoichiometry in oxygen sublattice on the electronic structure of $Na_{0.75}CoO_{2-y}$, calculated by KKR-CPA method. The combination of defects at two different O sites leads to

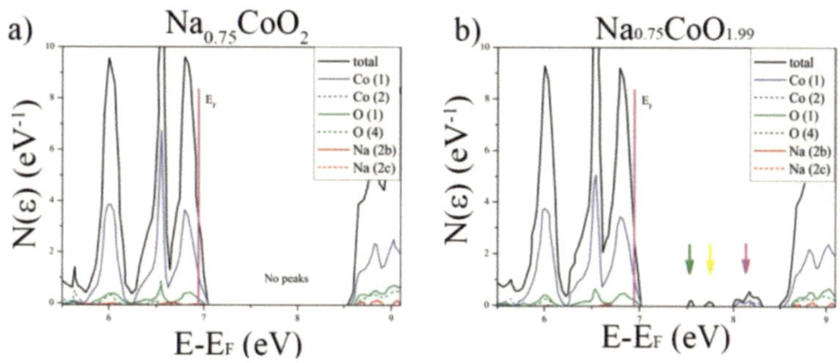

Figure 16.
KKR-CPA total and site-decomposed DOS for $Na_{0.75}CoO_2$ without oxygen vacancies (a) and nonstoichiometric $Na_{0.75}CoO_{1.99}$ (b).

three extra defect bands in the energy gap, formed essentially from 3d-Co and 2p-O states. To put this into context: for strictly stoichiometric $Na_{0.75}CoO_2$, there are no extra defect bands in the energy gap (**Figure 16a**).

Figure 17a presents an electronic model of a $Na/Na^+/Na_xCoO_{2-y}$ cell, with the proposed electronic structure of Na_xCoO_{2-y} (KKR-CPA calculations), which explains the step-like character of the discharge curve. In this structure, one can distinguish a valence band separated from the conductivity band by a 0.7 eV gap.

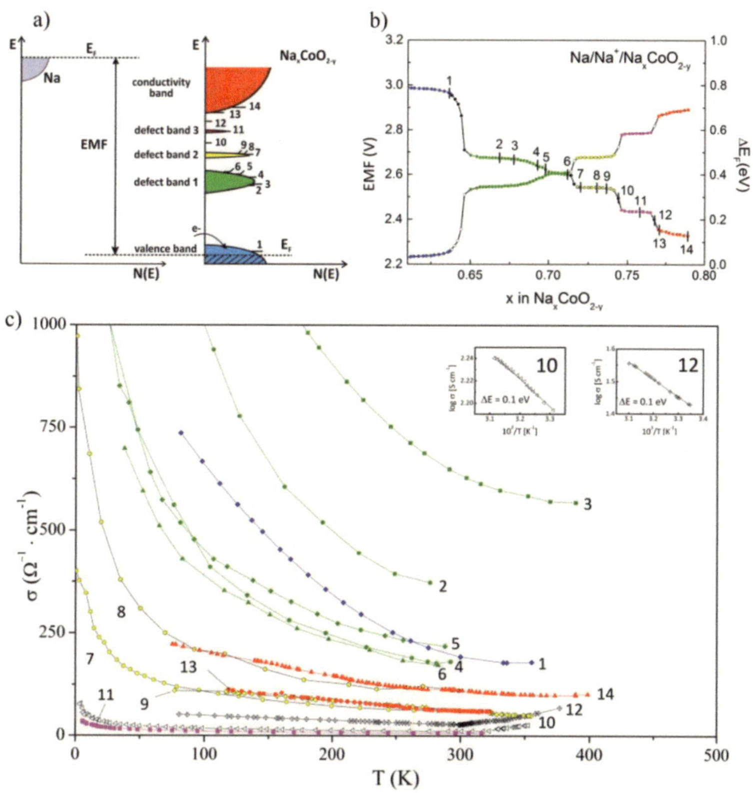

Figure 17.
*(a) Electronic diagram of a $Na/Na^+/Na_xCoO_{2-y}$ cell and a schematic diagram of the density of states for Na_xCoO_{2-y} calculated via KKR-CPA method [3, 32, 33] and measurements of transport properties. (b) Evolution of the Fermi level of Na_xCoO_{2-y} and electromotive force of a $Na/Na^+/Na_xCoO_{2-y}$ cell. (c) Temperature dependences of electrical conductivity of Na_xCoO_{2-y} at characteristic points of the discharge curve (corresponding to samples no. 1–14 in **Figure 17b**).*

Interestingly, three additional defect bands (denoted as "defect band 1," "defect band 2" and "defect band 3" in **Figure 17a**) are located within the band gap. Electrons introduced during sodium intercalation quickly fill the nearly fully-occupied valence band and then fill the three subsequent defect bands (**Figure 17a**); this is accompanied by an abrupt increase in the Fermi level and, in consequence, the step-like discharge curve (**Figure 17b**). Comprehensive studies of electrical conductivity, thermoelectric power, and electronic specific heat as well as oxygen K-edge NEXAFS, carried out at characteristic points of a $Na/Na^+/Na_xCoO_{2-y}$ cell's discharge curve, i.e. on "potential pseudo-plateaus" as well as potential jumps (**Figure 17b**), confirmed the anomalous, non-monotonic evolution of transport properties as a function of sodium content: metallic/semiconducting/metallic/semiconducting/metallic/semiconducting/metallic, which indicates anomalous, peaked density of states near the Fermi level and confirms the theoretical calculations of the electronic structure for this material. **Figure 17c** presents the measured temperature characteristics of electrical conductivity, documenting that a cathode material with a composition corresponding to a "potential pseudo-plateau" (e.g. samples no. 2–6), for which the Fermi level is located within a "defect band 1," exhibits metallic properties (cf. **Figure 17a–c**). Similarly, samples no. 7–9, at another "potential pseudo-plateau," also exhibit metallic behavior, although not as pronounced as the previous ones, due to the location of the Fermi level within the narrower "defect band 2" (**Figure 17a**). Sample no. 11, at the subsequent "potential pseudo-plateau," for which the Fermi level is located within the residual "defect band 3," exhibits even worse electrical conductivity, almost at the localization edge (**Figure 17a–c**). On the other hand, the cathode material with x_{Na} at potential jumps (samples no. 10 and 12, **Figure 17b**) corresponding to the location of the Fermi level in the energy gap of the density of states (**Figure 17a**), exhibits an activated character above 300 K, with an activation energy of 0.1 eV (**Figure 17c**, inset). This activation energy corresponds to the band gap between adequate bands in **Figure 17a** and the potential jumps of 0.1 V on the discharge curve (**Figure 17b**).

In order to support the electronic nature of step-like character of the discharge curve of a $Na/Na^+/Na_xCoO_{2-y}$ cell, electronic specific heat was measured for Na_xCoO_{2-y} at characteristic points of the cell's discharge curve, as in the case of the studies of electrical conductivity and thermoelectric power. The Sommerfeld coefficient, γ (directly proportional to density of states at the Fermi level, $\gamma = ⅓ \pi^2 k_B^2 N(E_F)$), determined at characteristic points of the cell's discharge curve, was highly consistent with electrical conductivity and thermoelectric power results, as depicted in **Figure 18** [4]. As expected, very low values of the Sommerfeld coefficient (very low density of states), of the order of 0.6–5 mJ mol^{-1} K^{-2}, were observed for samples for which the Fermi level is located in the energy gaps and in the residual "defect band 3" (denoted as H, I, J, and K in **Figure 18a** and **b**). Theoretical considerations had predicted values of 2–3 mJ mol^{-1} K^{-2}. The γ values for samples denoted B, C, D, and E, for which the Fermi level is located in "defect band 1," which exhibits a higher density of states, were of the order of 30 mJ mol^{-1} K^{-2}. The sample denoted as N, located in the conductivity band, with a high density of states, has a γ value of over 30 mJ mol^{-1} K^{-2}. The very low value of the Sommerfeld γ coefficient for the composition of $Na_{0.5}CoO_2$ (point A), for which a potential jump is recorded at discharge curve, corresponding to the location of the Fermi level in a band gap inside the modified valence band (**Figure 18a** and **b**), remains in very good agreement with the calculated electronic structure for this composition (**Figure 18c**), where the Fermi level is clearly located in the area of a diminishing density of states.

The above-presented new and globally unique results unambiguously document the "strictly electronic nature" of the step-like character of the discharge curve of a $Na/Na^+/Na_xCoO_{2-y}$ cell. The obtained results prove that the source of the step-like

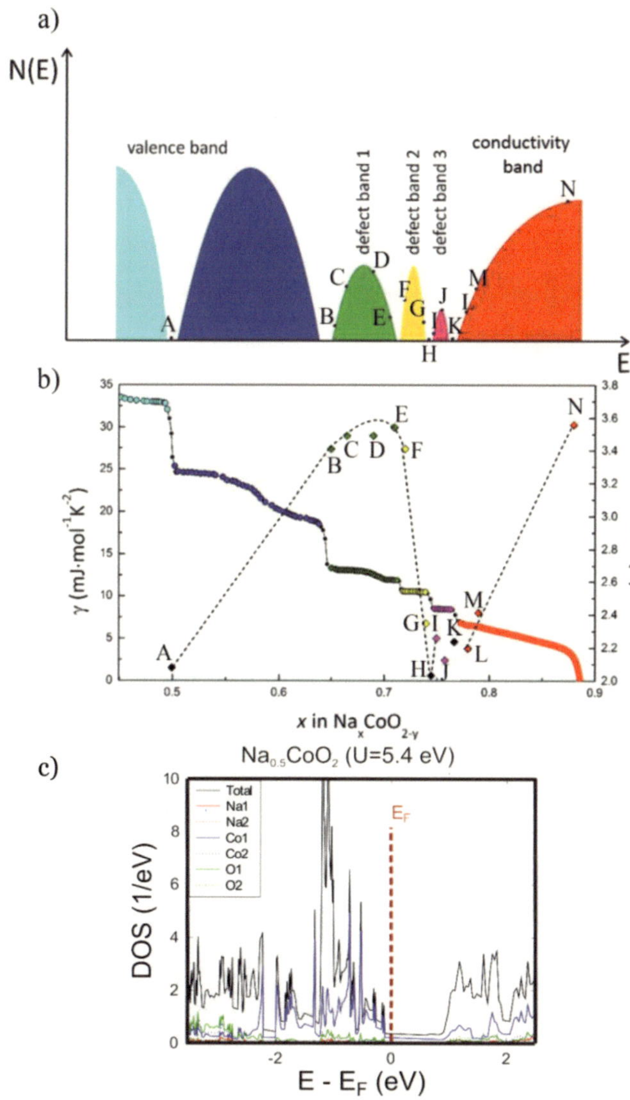

Figure 18.
(a) Density of states for Na_xCoO_{2-y} from KKR-CPA [3] electronic structure calculations and measurements of electrical conductivity and thermoelectric power, (b) Sommerfeld coefficient γ (points A-N) illustrating the density of states at the Fermi level in Na_xCoO_{2-y} plotted against the discharge curve [4]. (c) Electronic structure of $Na_{0.5}CoO_2$ from KKR-CPA computations corresponding to point A in figures (a) and (b), U—Hubbard energy.

discharge curve of a $Na/Na^+/Na_xCoO_{2-y}$ cell is the anomalous electronic structure of the Na_xCoO_{2-y} cathode material, induced by oxygen nonstoichiometry and two sodium crystallographic sites. They refute the current understanding of the principles of operation of lithium and sodium batteries, which assumes that mainly the ordering of the alkaline component and its mobility in the lithium or sodium sublattice plays a major role in the batteries parameters.

The author demonstrate that electronic structure "engineering" is a method suitable for controlling the properties of cathode materials by controlling its density of states and changing the character of the discharge curve from the unfavorable, step-like shape to the monotonic one via the modification of and control over the density of states function of a cathode material. The author has already obtained preliminary results that indicate the possibility of adjusting the properties of a

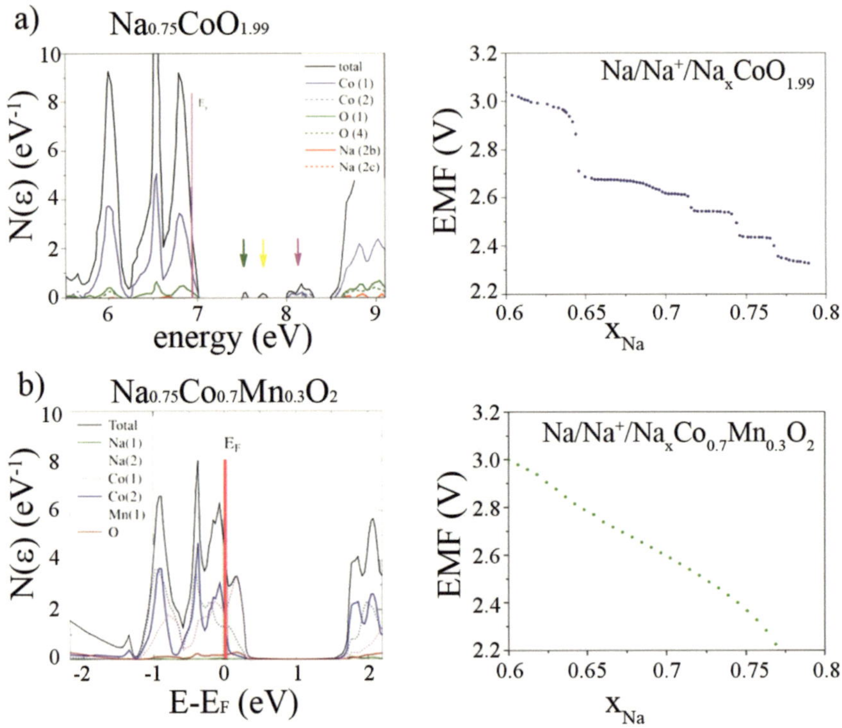

Figure 19.
(a) $Na_{0.75}CoO_{1.99}$ and step-like character of a discharge curve for $Na/Na^+/Na_xCoO_{1.99}$. (b) Density of states of $Na_{0.75}Co_{0.7}Mn_{0.3}O_2$ and a monotonic character of a discharge curve for $Na/Na^+/Na_xCo_{0.7}Mn_{0.3}O_2$ [42].

cathode material via the modification of the electronic structure of Na_xCoO_{2-y} performed by substitution of cobalt with manganese ($Na_xCo_{1-z}Mn_zO_{2-y}$). The electronic structure calculated for this system reveals a shift of the Fermi level deeper in the valence band; as a consequence of this shift, its evolution during the electrochemical deintercalation/intercalation of sodium takes place within the valence band, resulting in a monotonic, favorable character of the discharge/charge curve for an $Na/Na^+/Na_xCo_{1-z}Mn_zO_{2-y}$ cell (**Figure 19**).

Acknowledgements

The project was funded by the National Science Centre Poland under UMO-2015/19/B/ST8/00856 and UMO-2016/23/B/ST8/00199 grants and Polish Ministry of Science and Higher Education under project AGH No. 11.11.210.911. This work was carried out using infrastructure of the Laboratory of Materials for Renewable Energy Conversion and Storage, Centre of Energy AGH.

Author details

Janina Molenda
Department of Hydrogen Energy, Faculty of Energy and Fuels, AGH University of Science and Technology, Krakow, Poland

*Address all correspondence to: molenda@agh.edu.pl

References

[1] Than D, Molenda J, Stokłosa A. Correlation between electronic and electrochemical properties of $Li_xTi_{1+y}S_2$. Electrochimica Acta. 1991;**36**:1555-1560

[2] Molenda J, Stokłosa A, Bąk T. Modification in the electronic structure of cobalt bronze Li_xCoO_2 and the resulting electrochemical properties. Solid State Ionics. 1989;**36**:53-58

[3] Molenda J, Baster D, Molenda M, Swierczek K, Tobola J. Anomaly in the electronic structure of the Na_xCoO_{2-y} cathode as a source of its step-like discharge curve. Physical Chemistry Chemical Physics. 2014;**16**:14845-14857

[4] Molenda J, Baster D, Gutowska MU, Szewczyk A, Puźniak R, Tobola J. Electronic origin of the step-like character of the discharge curve for Na_xCoO_{2-y} cathode. Functional Materials Letters. 2014;**7**

[5] Molenda J, Delmas C, Hagenmuller P. Electronic and electrochemical properties of Na_xCoO_{2-y} cathode. Solid State Ionics. 1983;**9**:431-435

[6] Molenda J, Baster D, Milewska A, Świerczek K, Bora DK, Braun A, et al. Electronic origin of difference in discharge curve between Li_xCoO_2 and Na_xCoO_2 cathodes. Solid State Ionics. 2015;**271**:15-27

[7] Molenda J, Stokłosa A. Electronic structure and electrochemical properties of VO_2. Solid State Ionics. 1989;**36**:43-52

[8] Gędziorowski B, Kondracki Ł, Świerczek K, Molenda J. Structural and transport properties of $Li_{1+x}V_{1-x}O_2$ anode materials for Li-ion batteries. Solid State Ionics. 2014;**262**

[9] Molenda J, Bąk T. Effect of tungsten on the electrical and electrochemical properties of the $Li_yW_xVO_2$ cathode. Physica Status Solidi. 1993;**135**:263-271

[10] Molenda J, Wilk P, Marzec J. Structural, electrical and electrochemical properties of $LiNiO_2$. Solid State Ionics. 2002;**146**:73-79

[11] Molenda J, Kubik A. Transport properties and reactivity of tungsten trioxide. Solid State Ionics. 1999;**117**:57-64

[12] Molenda J, Stokłosa A, Bak T. Transport properties of $YBa_2Cu_3O_{7-y}$ at high temperatures. Physica C: Superconductivity and Its Applications. 1991;**175**:555-565

[13] Molenda J, Bak T, Stokosa A. Influence of lithium on the electronic structure of $YBa_2Cu_3O_{7-\delta}$. Physica C: Superconductivity and Its Applications. 1993;**207**:147-158

[14] Molenda J, Marzec J, Świerczek K, Ojczyk W, Ziemnicki M, Molenda M, et al. The effect of 3d substitutions in the manganese sublattice on the charge transport mechanism and electrochemical properties of manganese spinel. Solid State Ionics. 2004;**171**:215-227

[15] Molenda J, Marzec J, Świerczek K, Pałubiak D, Ojczyk W, Ziemnicki M, et al. The effect of 3d substitutions in the manganese sublattice on the electrical and electrochemical properties of manganese spinel. Solid State Ionics. 2004;**175**:297-304

[16] Molenda J, Marzec J. Functional Cathode Materials For Li-ion Batteries - Part III: Potential Cathode Materials $Li_xNi_{1-y-z}Co_y Mn_zO_2$ and $LiMn_2O_4$. Functional Materials Letters. 2009;**2**:1-7

[17] Milewska A, Kondracki L, Molenda M, Bakierska M, Molenda J. Structural, transport and electrochemical properties of $LiNi_{0.5-y}CuyMn_{1.5}O_{4-\delta}$ spinel cathode materials. Solid State Ionics. 2014;**267**:27-31

[18] Molenda J. Electronic Processes in Electrode Materials of A_xMX_2-Type. Physica Status Solidi. 1984;**122**:591-598

[19] Molenda J. Material problems and prospects of Li-ion batteries for vehicles applications. Functional Materials Letters. 2011;**4**:107-112

[20] Milewska A, Świerczek K, Tobola J, Boudoire F, Hu Y, Bora DK, et al. The nature of the nonmetal–metal transition in Li_xCoO_2 oxide. Solid State Ionics. 2014;**263**:110-118

[21] Molenda J, Milewska A, Zając W, Rybski M, Tobola J. Correlation between electronic structure, transport and electrochemical properties of a $LiNi_{1-y-z}Co_yMn_zO_2$ cathode material. Physical Chemistry Chemical Physics. 2017;**19**:25697-25706

[22] Belov D, Yang M-H. Failure mechanism of Li-ion battery at overcharge conditions. Journal of Solid State Electrochemistry. 2008;**12**:885-894

[23] Maleki H, Deng G, Anani A, Howard J. Thermal stability studies of Li-Ion cells and components. Journal of the Electrochemical Society. 1999;**146**:3224-3229

[24] Dahn JR, Fuller EW, Obrovac M, Von Sacken U. Thermal stability of Li_xCoO_2, Li_xNiO_2 and λ-MnO_2 and consequences for the safety of Li-ion cells. Solid State Ionics. 1994;**69**:265-270

[25] Chen Z, Lu Z, Dahn JR. Staging phase transitions in Li_xCoO_2. Journal of the Electrochemical Society. 2002;**149**:A1604-A1609

[26] Van der Ven A, Aydinol MK, Ceder G, Kresse G, Hafner J. First-principles investigation of phase stability in Li_xCoO_2. Physical Review B. 1998;**58**:2975

[27] Marianetti CA, Kotliar G, Ceder G. A first-order Mott transition in Li_xCoO_2. Nature Materials. 2004;**3**:627-631

[28] Ménétrier M, Saadoune I, Levasseur S, Delmas C. The insulator-metal transition upon lithium deintercalation from $LiCoO_2$: Electronic properties and ^7Li NMR study. Journal of Materials Chemistry. 1999;**9**:1135-1140

[29] Ohzuku T, Ueda A. Solid-State redox reactions of $LiCoO_2$ ($R\bar{3}m$) for 4 volt secondary lithium cells. Journal of the Electrochemical Society. 1994;**141**:2972-2977

[30] Mizushima K, Jones PC, Wiseman PJ, Goodenough JB. Li_xCoO_2 (0< x<-1): A new cathode material for batteries of high energy density. Materials Research Bulletin. 1980;**15**:783-789

[31] Chebiam RV, Kannan AM, Prado F, Manthiram A. Comparison of the chemical stability of the high energy density cathodes of lithium-ion batteries. Electrochemistry Communications. 2001;**3**:624-627

[32] Kaprzyk S, Bansil A. Green's function and a generalized Lloyd formula for the density of states in disordered muffin-tin alloys. Physical Review B. 1990;**42**:7358-7362

[33] Stopa T, Kaprzyk S, Tobola J. Linear aspects of the Korringa–Kohn–Rostoker formalism. Journal of Physics. Condensed Matter. 2004;**16**:4921-4933

[34] Bansil A, Kaprzyk S, Mijnarends PE, Tobola J. Electronic structure and magnetism of $Fe_{[3-x]}V_{[x]}X$ (X = Si, Ga, and Al) alloys by the KKR-CPA method. Physical Review B. 1999;**60**:396-412

[35] Berthelot R, Carlier D, Delmas C. Electrochemical investigation of the P2–Na_xCoO_2 phase diagram. Nature Materials. 2011;**10**:74-80

[36] Meng YS, Hinuma Y, Ceder G. An investigation of the sodium patterning in Na_xCoO_2 (0.5 < = x < =1) by density functional theory methods. The Journal of Chemical Physics. 2008;**128**:104708

[37] Hinuma Y, Meng YS, Ceder G. Temperature-concentration phase diagram of P2-Na_xCoO_2 from first-principles calculations. Physical Review B. 2008;**77**:224111

[38] Lee M, Viciu L, Li L, Wang Y, Foo ML, Watauchi S, et al. Large enhancement of the thermopower in Na_xCoO_2 at high na doping. Nature Materials. 2006;**5**:537-540

[39] Takada K, Sakurai H, Takayama-Muromachi E, Izumi F, Dilanian RA, Sasaki T. Superconductivity in two-dimensional CoO_2 layers. Nature. 2003;**422**:53-55

[40] Banobre-Lopez M, Rivadulla F, Caudillo R, López-Quintela MA, Rivas J, Goodenough JB. Role of doping and dimensionality in the superconductivity of Na_xCoO_2. Chemistry of Materials. 2005;**17**:1965-1968

[41] Hertz JT, Huang Q, McQueen T, Klimczuk T, Bos JWG, Viciu L, et al. Magnetism and structure of Li_xCoO_2 and comparison to Na_xCoO_2. Physical Review B. 2008;**77**:75119

[42] Baster D, Zajac W, Kondracki Ł, Hartman F, Molenda J. Improvement of electrochemical performance of $Na_{0.7}Co_{1-y}Mn_yO_{2-\delta}$ cathode material for rechargeable sodium-ion batteries. Solid State Ionics. 2016;**5**:213

Chapter 4

Cathode for Thin-Film Lithium-Ion Batteries

Yuan-Li Ding

Abstract

Thin-film lithium-ion batteries (LIBs) have attracted considerable attention for energy storage device application owing to their high specific energy compared to conventional LIBs. However, the significant breakthroughs of electrochemical performance for electrode materials, electrolyte, and electrode/electrolyte interface are still highly desirable. This chapter firstly gives an overview of cathode materials including lithium-containing cathode (e.g., $LiCoO_2$, $LiMn_2O_4$, $LiFePO_4$, $LiNi_{1-x-y}Mn_xCo_yO_2$, $LiNi_{0.5}Mn_{1.5}O_4$) and lithium-free cathode (e.g., vanadium oxides) for LIBs in terms of specific capacity, energy density, working voltage, cycling life, and safety. In the meanwhile, the existing drawbacks and limitations of various battery chemistries are also analyzed. Furthermore, some modification strategies for these cathode materials have also been discussed for improving electrochemical performance. Finally, the thin-film Li-ion battery applications of these cathode materials are summed up toward next-generation flexible and high-energy devices.

Keywords: cathode, nanotechnology, energy density, thin-film lithium-ion batteries

1. Overview of cathode materials

Since lithium-ion batteries (LIBs) gained the commercial success in 1991 [1], they have received considerable attention in various areas including consumer electronics, power tools, electric vehicles, and grid energy storage owing to their high energy density, low cost, long cycle life, and environmentally benignity. The energy density of LIBs in 1991 is only 80 Wh kg^{-1} at a cell level and has been currently increased to 200–250 Wh kg^{-1} (at a cell level). The main reason is attributed to the significant improvement of battery materials and battery technology. Especially for cathode materials, they dominate the whole cell performance compared to graphite anode.

Up to now, various cathode materials have been proposed and developed such as layered structure $LiCoO_2$, $LiMnO_2$, $LiNiO_2$, ternary $LiNi_{1-x-y}Mn_xCo_yO_2$, spinel $LiMn_2O_4$ (LMO), spinel $LiNi_{0.5}Mn_{1.5}O_4$ (LNMO), and polyanion-based cathode including $LiFePO_4$ and $Li_3V_2(PO_4)_3$ [2], as shown in **Figure 1**. Typically, transition metals of such cathode materials undergo an oxidation process to higher oxidation state when Li^+ ions are removed. While oxidation of the transition metals can retain charge neutrality in the compound, Li^+ removal/insertion usually causes phase changes and structural strain. In this regard, cathode materials that are stable enough over a wide composition range must be employed. Thus, the structural stability of cathode materials is a key during Li^+ extraction/insertion processes. As for discharge process, Li^+ ions are reinserted into the cathode materials and electrons from anode reduce transition metal ions in the cathode to a lower valence. The rates

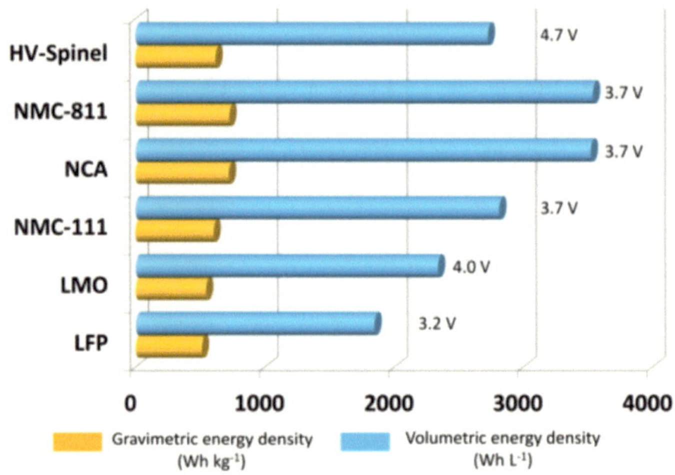

Figure 1.
Gravimetric and volumetric energy densities of various cathode materials at a material level. LFP, $LiFePO_4$; LMO, $LiMn_2O_4$; NMC-111, $LiNi_{1/3}Mn_{1/3}Co_{1/3}O_2$; NCA, $LiNi_{0.8}Co_{0.15}Al_{0.05}O_2$; NMC-811, $LiNi_{0.8}Mn_{0.1}Co_{0.1}O_2$; HV-spinel, $LiNi_{0.5}Mn_{1.5}O_4$.

of the two processes, as well as access of Li^+ ions in the electrolyte to electrode surface, dominate the maximum discharge current. Therefore, cathode performance depends critically on the microstructure and morphology of electrode as well as the intrinsic electrochemical properties of the cathode materials. For instance, much effort has been currently focused on the development of nanosized electrode materials for shortening Li^+/electron transport length and increasing contact areas of electrode/electrolyte interface. This chapter will mainly discuss cathode materials for LIBs and their potential applications for thin-film Li-ion microbatteries.

2. Layered transition metal oxide cathode

Layered transition metal oxides are the most commonly used cathode materials for LIBs such as $LiCoO_2$, $LiNiO_2$, $LiMnO_2$, and $LiTiS_2$. Such cathode materials have typical layered structure, as illustrated in **Figure 2**. $LiCoO_2$ is relatively one of the most mature cathode materials, which will be discussed in the following section. However, the crystal structure of Li_xCoO_2 becomes unstable when Li^+ extracts beyond $x > 0.5$, leading to oxygen release [3]. Considering the shortcomings and costly Co resource of $LiCoO_2$, layered ternary transition metal oxides $LiNi_{1-x-y}Mn_xCo_yO_2$ (NMC) have been widely investigated and developed in recent years.

2.1 $LiCoO_2$

As one of the typical intercalation compound families, $LiCoO_2$ was first reported and studied by John B. Goodenough in 1980 for LIBs [4]. It has an a-$NaFeO_2$ structure containing only one kind of cation in the transition meta layer. $LiCoO_2$ is widely used as cathode materials for LIBs, especially for portable electronics. Although $LiCoO_2$ is the most widely used cathode materials for LIBs, the practical capacity available is only around 140 mAh g^{-1}, half of the theoretical capacity (ca. 280 mAh g^{-1}). It corresponds to the 50% Li removal from $LiCoO_2$ because further lithium extraction from $LiCoO_2$ usually induces the structural transition from hexagonal to monoclinic phase and Co dissolution at the highly oxidized state, thus leading to rapid capacity decay. To enhance the capacity retention of $LiCoO_2$ at the

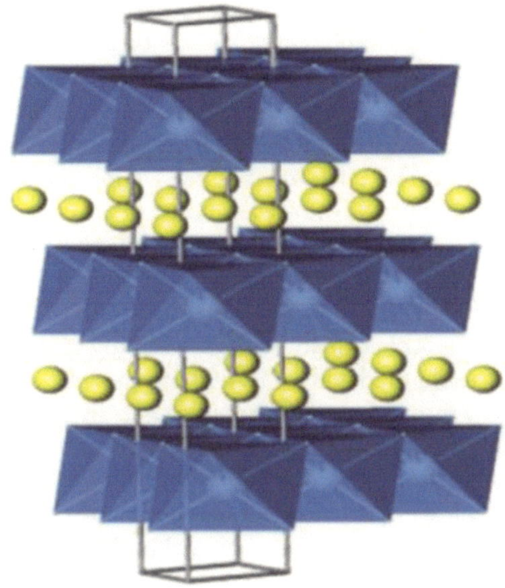

Figure 2.
Layered structure of $LiMO_2$ (M = Ni, Co, Mn), showing the lithium ions between the transition metal oxides.

higher cutoff potential (4.4–4.5 V), surface modifications (e.g., surface coating) on the $LiCoO_2$ surface have been employed since such the strategies can effectively restrict the lattice constant change and the transition of hexagonal to monoclinic phase change upon cycling. The mechanism of such capacity improvement for $LiCoO_2$ was not widely identified. Dahn group found that the poor capacity retention of $LiCoO_2$ at the high delithiation potential (4.5 V) was attributed to the growth of impedance at the $LiCoO_2$ surface caused by side reactions [5]. As for $LiCoO_2$, although a higher reversible capacity by using a higher cutoff potential can be obtained, such a cathode still faces a thermal stability concern, resulting in a safety issue. Moreover, Co in the delithiated $Li_{1-x}CoO_2$ possesses high oxidizing power and easily induces the more side reactions with organic electrolytes. Thus, $LiCoO_2$ is mainly used as low-power devices such as consumer electronics.

2.2 $LiNi_{1-x-y}Mn_xCo_yO_2$

Although $LiCoO_2$ exhibits the most early commercial success, the intrinsic structural stability upon delithiation and high raw material cost of $LiCoO_2$ have pushed the development of a new class cathode materials. As an alternative cathode, layered structured Li_xNiO_2 has been developed owing to its relatively low cost and high working potential [6]. However, pristine Li_xNiO_2 is not a promising cathode candidate because of its some drawbacks. For example, more than 200 mAh g^{-1} can be obtained from stoichiometric $LiNiO_2$ at a relatively high voltage (4.5 V vs. Li/Li^+). Moreover, a 20% capacity loss occurs in the first cycle owing to the structural modification when charged below x = 0.5. Besides electrochemical properties, stoichiometric $LiNiO_2$ with a Li/Ni ratio of 1:1 is still difficult to be synthesized because of the cation mixing of Ni^{2+} and Ni^{3+} owing to the similarity of ionic radii of Li^+ (0.76 Å) and Ni^{2+} (0.69 Å) [7]. Moreover, Li_xNiO_2 is usually Li-deficient because it is difficult to prepare stoichiometric $LiNiO_2$. To address this issue, the synthesis condition usually needs oxygen instead of air atmosphere and a slight excess of $LiNO_3$.

Considering the advantages and shortcomings of $LiNiO_2$, metal-doped $LiNiO_2$, especially ternary layered cathode NMC and $LiNi_{0.8}Co_{0.15}Al_{0.05}O_2$ (NCA), has recently attracted considerable attention because it shows good structural stability, higher specific capacity up to 220 mAh g^{-1}, higher energy density, and relatively simple synthesis [8]. By introducing Mn and Co or Al and Co into $LiNiO_2$, the cation mixing of Ni^{2+}/Ni^{3+} can be effectively restricted. In the meanwhile, $LiNi_{1-x-y}Mn_xCo_yO_2$ can be synthesized in the air without the necessity to use pure oxygen atmosphere. As for NMC cathode, $LiNi_{1/3}Mn_{1/3}Co_{1/3}O_2$ (NMC-111) is currently the most common form of NMC cathode materials and is widely employed in LIBs [9]. To further enhance energy density and reduce cost, increasing Ni content and lowering Co content in NMC cathode are effective solutions such as $LiNi_{0.5}Mn_{0.3}Co_{0.2}O_2$ (NMC-532), $LiNi_{0.6}Mn_{0.2}Co_{0.2}O_2$ (NMC-622), and $LiNi_{0.8}Mn_{0.1}Co_{0.1}O_2$ (NMC-811), as shown in **Figure 3**.

It is well known that Ni is beneficial for achieving high energy density but poor stability, especially at the charged state. Mn will induce the formation of a spinel structure to realize a low internal resistance but results in a low energy density. Compared to Ni and Mn, Co has the limited resources (Co reserve available is around 7.1 million tons) [10], high price, and high toxicity. Thus, NMC battery chemistry will gradually lessen the use of Co element and simultaneously employ high Ni and low Mn for realizing the best electrochemical performance by combining the respective merits of Ni, Mn, and Co. NMC-111, NMC-442, and NMC-532 are currently the state of the art of cathode materials. In the very recently, Ni-rich NMC cathodes including NMC-622 and NMC-811 have been widely investigated and will be adopted in the LIBs owing to their higher energy density and lower cost. This is very favorable for automotive applications. On the other hand, Mn^{3+} has the lowest octahedral site stabilization energy (OSSE) and tends to migrate, resulting in a layered-to-spinel transition upon lithiation/delithiation process. In contrast, Co^{3+} has the highest OSSE and exhibits excellent structural stability, but it suffers from poor chemical stability during lithium extraction over 50%, which is attributed to an overlap of the t_{2g} band of low-spin $Co^{3+}/^{4+}$ with the top of the 2p band of O^{2-} [11], as shown in **Figure 3b**. Compared to Co, Mn provides better chemical stability because $Mn^{3+}/^{4+}$ e_g band is further above the top of the 2p band of the O^{2-}. Compared to Co and Mn, Ni shows a moderate structural and chemical stability considering that it has a higher OSSE than Mn^{3+}, and the e_g band of low-spin $Ni^{3+}/^{4+}$ just touches the top of the O^{2-} 2p band. Thus, developing Co-less and Ni-rich NMC cathode will be very desirable.

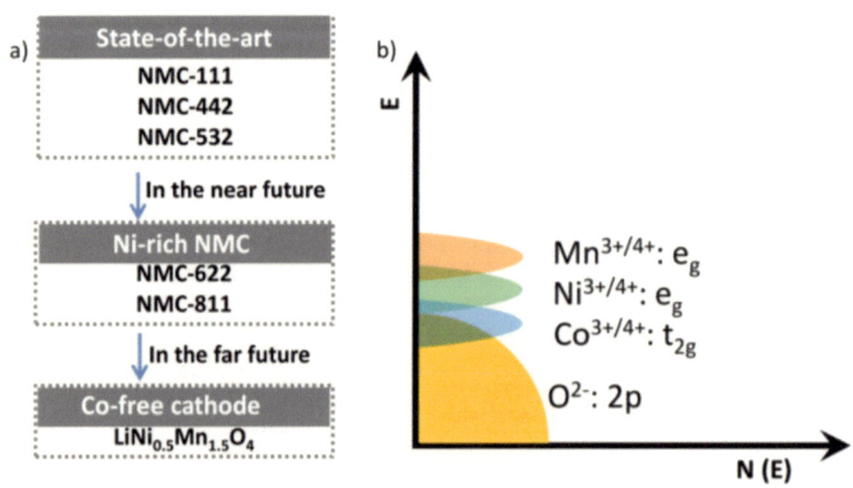

Figure 3.
(a) The state-of-the-art NMC cathode materials, near-future NMC materials, and far-future cathode materials. (b) Schematic diagram of the positions of the various redox couples relative to the top of the oxygen: 2p band.

The commercial success of NMC-111 facilitates the development of other NMC materials by combining the individual advantages of Ni, Mn, and Co in the layered crystal structure with varying concentration to realize high capacity and high energy density. As discussed previously, a high Ni content is favorable to achieve a high reversible capacity. On the other hand, the introduction of Mn can enhance the structural and thermal stability at the deeply delithiated state, while Co dopant in NMC can improve the layered ordering as well as rate performance and specific capacity owing to its redox activity. Sun et al. have systematically investigated the effect of chemical compositions for the ternary NMC materials on reversible capacity, cyclability, and thermal stability [12], as displayed in **Figure 4**. The NMC materials with higher Ni content show higher reversible capacities, but it shows a decreased capacity retention at both 25 and 55°C. In addition, the exothermic reaction for Ni-rich NMC usually occurs at a lower temperature with a larger heat generation upon charging to 4.3 V vs. Li/Li^+, which is a good agreement with the phenomena of capacity retention decrease with increasing Ni content (**Figure 4**). Such findings confirm that Ni-rich NMC possesses a higher specific capacity but poor thermal stability and capacity retention upon cycling, as shown in **Figure 4**. To address this problem, a novel material design by tuning chemical composition in the core and shell has been proposed by Sun group [13]. The first generation is a core-shelled structure based on Ni-rich NMC-811 as core and Mn-rich $LiNi_{0.5}Mn_{0.5}O_2$ as shell (**Figure 5a**). The core can offer high capacity, while the shell guarantees the surface stability. Such structural design, however, also has some drawbacks. For example, boundary cracks between active particles often formed owing to different

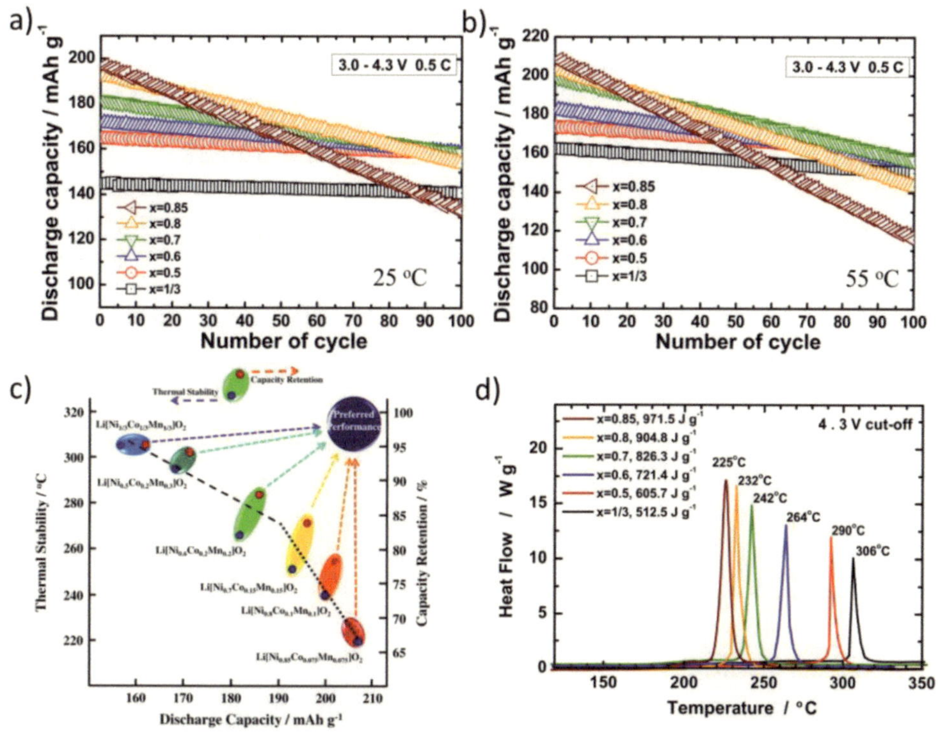

Figure 4.
Discharge capacity as a function of cycle number for various LiNixMnyCozO2 (x = 1/3, 0.5, 0.6, 0.7, 0.8, and 0.85) cathodes at (a) 25°C and (b) 55°C. The current density is 100 mA g^{-1} (0.5 C) in the voltage range of 3.0–4.3 V [12]. (c) The map of relationship between discharge capacity and thermal stability and capacity retention for $LiNi_xMn_yCo_zO_2$ (x = 1/3, 0.5, 0.6, 0.7, 0.8, and 0.85) cathode. (d) DSC curves of the above NMC cathode materials.

volume changes between the core and the shell structures upon cycling tend to induce the formation of boundary cracks, leading to major mechanical fracture of active particles and capacity decay. Thus, another core-shelled gradient-type and full concentration gradient-type NMC materials have been proposed and developed [14], as illustrated in **Figure 5b**. This unique design minimizes the propagation of boundary cracks and is beneficial to enhancing cycling stability.

2.3 Layered NCA cathode

Similar to NMC-811 cathode, $LiNi_{0.8}Co_{0.15}Al_{0.05}$ (NCA) is also a Ni-rich cathode, showing a relatively high energy density and low cost. The introduction of element Al in NCA can minimize the detrimental phase transition and improve the thermal behavior of the cathode. Al is a favorable dopant in a specific capacity because of its lower atomic weight than the other transition metal elements. Thus, Al-doped $LiNiO_2$ shows an increased specific capacity. Moreover, the presence of Al^{3+} in the transition metal layer tends to reduce the a-axis but increase the c-axis parameters because of α-$LiAlO_2$ (a = 2.8 Å and c = 14.23 Å), which decrease the cationic disorder in the Li layer. It is similar to the effect of Co^{3+}. However, a large amount of Al doping would lead to the reduction of specific capacity because Al is electrochemically inactive. On the other hand, Al^{3+} can effectively enhance the working voltage owing to the weakening of the Ni-O bond by the stronger Al-O bond through the inductive effect. Thus, a reasonable balance between reduced capacity and increased working voltage should be considered for designing high-performance Al-doped $LiNiO_2$. It is acknowledged that Al doping (5 atom%) and Co doping (15 atom%) is the optimized results [11]. In recent years, double metal doping has been widely accepted to improve the structural stability and electrochemical performance of layer transition metal oxides.

As discussed previously, the surface stability of Ni-rich cathode is a key issue for their electrochemical performance. The residual lithium oxide at the particle surface after annealing tends to absorb moisture and CO_2 in air during storage, leading to the formation of some unwanted products on the surface such as LiOH and Li_2CO_3. These products would react with electrolyte, forming surface interface. In addition, Ni^{4+} at the delithiated state would accelerate the decomposition of electrolyte, further resulting in the thickening of interfaces and sluggish Li^+ diffusion kinetics. Considering these problems, surface protection of Ni-rich cathode materials is highly desirable such as various surface coatings including metal oxide, phosphates, carbon, fluorides, artificial SEI, and conducting polymers [11].

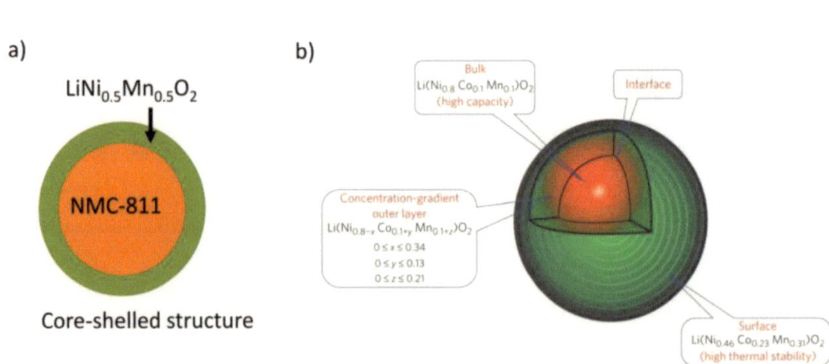

Figure 5.
(a) Core-shelled structure NMC: core: NMC-811 and shell: LiNi0.5Mn0.5O2; (b) concentration gradient-based NMC cathode.

3. Spinel cathode

3.1 Spinel $LiMn_2O_4$

In contrast to layered $LiCoO_2$, spinel $LiMn_2O_4$ possesses more stable structure and higher voltage (4.0 V). As shown in **Figure 6**, $LiMn_2O_4$ has a cubic crystal structure with Li occupying the tetrahedral 8a sites and Mn occupying the octahedral 16d sites. The interstitial sites for Li, 16c, share faces with the occupied sites, forming a continuous three-dimensional diffusion network by the [Mn_2O_4] framework. Thackeray et al. have demonstrated that lithium can be reversibly intercalated and deintercalated for the range of $0 < x < 2.2$ in $Li_xMn_2O_4$ in the cutoff range (2–4.5 V) with two plateaus at 4.0 and 3.0 V [15]. However, $LiMn_2O_4$ is usually cycled in the voltage range of 3.5–4.5 V because of the poor structural stability during cubic-tetragonal phase transition for $1 < x < 2$ in the cutoff window (2–4.5 V). Such cubic-tetragonal phase transition is driven by the Jahn-Teller distortion of MnO_6 octahedron as Mn^{4+} is reduced to Mn^{3+}, leading to a 6.5% increase of the unit cell volume. Thus, the specific capacity available for $LiMn_2O_4$ is theoretical 148 mAh g^{-1}. More importantly, $LiMn_2O_4$ just needs to use earth-abundant and low-cost Mn resources without using Ni and Co element, which is very beneficial for its large-scale application.

Although spinel $LiMn_2O_4$ shows attractable merits for LIBs, it also suffers from some drawbacks, such as Jahn-Teller distortion and Mn dissolution. To inhibit the Jahn-Teller distortion, doping with Li or multivalent cations (e.g., Al, Mg) has been widely used because partial replacement of Mn can reduce the Mn^{3+} concentration and increase the average valence of Mn, thus suppressing the onset of Jahn-Teller effect. Although metal-doped $LiMn_2O_4$ shows the enhanced electrochemical properties, it inevitably reduces the theoretical capacity owing to inactive element introduction. In this regard, a small amount of metal substitution is desirable. Another problem for spinel $LiMn_2O_4$ is the Mn dissolution owing to the disproportionation reaction of Mn^{3+} by forming Mn^{4+} and soluble Mn^{2+} catalyzed by acidic species in the electrolyte. Moreover, Mn^{2+} can react with $LiPF_6$ salt to produce more acidic by-products. To mitigate or address such problems, the surface modification strategy has been extensively employed.

3.2 High-voltage spinel $LiNi_{0.5}Mn_{1.5}O_4$

Besides 4 V-spinel, another Mn-based spinel $LiNi_{0.5}Mn_{1.5}O_4$ (LNMO) also shows considerable attention for LIBs in recent years. Owing to high working voltage (4.7 V vs. Li/Li^+), such spinel shows the highest specific energy (650 wh kg^{-1} at a material level) compared to other cathode materials (LCO, 540 Wh kg^{-1}; LFP,

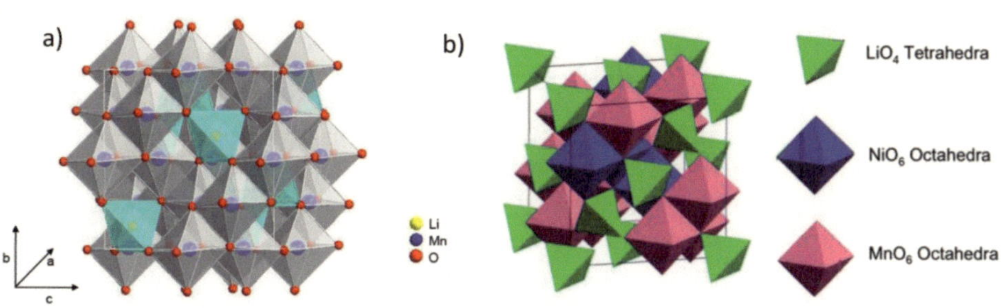

Figure 6.
Crystal structures of (a) spinel LiMn2O4 and (b) spinel LiNio.5Mn1.5O4.

500 Wh kg^{-1}; LMO, 500 Wh kg^{-1}). The interstitial space in the spinel framework offers 3D Li^+ diffusion channels. Owing to its high working voltage, LNMO has become a very promising candidate for replacing spinel $LiMn_2O_4$ cathode. Moreover, relatively benign metal elements make such a material a top choice for next-generation high-power batteries. However, there are still some serious limitations to restrict its commercial applications. Ni is active with the two Ni^{2+}/Ni^{3+} and Ni^{3+}/Ni^{4+} couples at a 4.7 V, while Mn is inactive without capacity contribution. Although Ni^{2+}/Ni^{3+} and Ni^{3+}/Ni^{4+} electrochemical couples provide favorable high working voltage, they also bring some unwanted side reactions between active component and electrolyte, which gives rise to a considerable thickening of the protective solid-electrolyte interphase (SEI) layer passivating the material surface, thus leading to Li^+ ion diffusion obstacle.

High redox couple of Ni^{3+}/Ni^{4+} easily incurs the decomposition of conventional carbonate-based electrolyte. It is difficult for pure LNMO phase to be synthesized because of mixed ordered and disordered LNMO phase owing to oxygen deficiency caused by high-temperature annealing. In addition, the $Li_xNi_{1-x}O_2$ impurity probably forms as the second phase. Compared to ordered LNMO phase with $P4_332$ space group, disordered phase is relatively easily synthesized. The disordered spinel shows a better electrochemical performance than ordered spinel because the disordered one has an increased Li^+ ion diffusion coefficient. However, the presence of a small amount Mn^{3+} in the disorder phase will cause a non-negligible capacity fading during cycling, especially at elevated temperature. In this respect, introducing metal doping is an effective strategy to increase the average valence of Mn in the spinel. Mn^{3+} concentration in the spinel can be also decreased by reannealing treatment, which can also reduce oxygen defect and impurity amount, resulting in better electrochemical performance. In recent years, various strategies have been widely used to enhance electrochemical performance of LNMO, such as doping and surface coating. Metal doping can provide better chemical and structural stability and higher electrical conductivity. For example, Ti-doped LNMO improved the disordering of the transition metals and lowered the symmetry from simple cubic structure ($P4_332$) to face-centered spinel ($Fd3m$) [16]. Importantly, Ti-doped LNMO spinel shows higher working voltage, faster Li^+ ion diffusion, and better rate capability than undoped spinel. Fe doping in the high-voltage spinel can stabilize the solid during repeated cycling, leading to better cycling performance [17]. Thus, double doping by using Ti and Fe is an effective strategy to synergistically enhance the structural stability of LNMO. Additionally, other dopants including Cr, Ru, Zr, Al, and Mg can show better rate capability and cycling performance. Besides metal doping, anion dopant can also enhance electrochemical performance of high-voltage spinel. For example, the F-doped LNMO can restrict the formation of NiO impurity and change the lattice parameters and bonding energy [18], leading to fine structure and significantly enhanced cycling performance.

On the other hand, similar to layered NMC, the interface stability of LNMO between electrode and electrolyte is also a key for achieving long-term cycling performance. It has been demonstrated that better stability can be obtained by coating more stable surface layer, such as metal (Zn, Au, Ag), metal oxides (ZnO, SnO_2, ZrO_2), and metal phosphates (Li_3PO_4). For instance, ZnO-coated LNMO can effectively suppress Mn dissolution and increase the stability. More importantly, HF content in the electrolyte can be restricted because ZnO protection layer acts as a scavenger of fluoride anions from HF generated from the decomposition of $LiPF_6$ salt in the electrolyte by transforming HF to ZnF_2 [19, 20]. Li_3PO_4 protects the surface of LNMO and functions as a solid-electrolyte interface between active component and solid polymer electrolyte (SPE) to prevent the SPE degradation [21]. Such findings show that surface coating can not only enhance structural stability of LNMO but also restrict Mn dissolution, thus leading to improved cycling stability.

4. Polyanion-based cathode

4.1 Olive $LiFePO_4$

Polyanion cathode materials have received considerable attention in LIBs owing to their stable crystal structure and long-term cycling performance. Since John B. Goodenough proposed $LiFePO_4$ as cathode materials for LIBs in 1997, it has received much investigation including academic and industrial activities. As shown in **Figure 7**, $LiFePO_4$ possesses an olive structure with space group Pnma. The O atoms are located in a slightly distorted, hexagonally close-packed arrangement. The P atoms occupy tetrahedral sites, while Li and Fe atoms occupy octahedral 4c and 4a sites, respectively. The LiO_6 octahedra from edge-sharing chain along the b-axis. Each FeO_6 octahedron is linked with four FeO_6 octahedra by common corners in the b–c plane, forming zigzag planes. One FeO_6 octahedron has common edges with two LiO_6 octahedra. PO_4 groups share one edge with one FeO_6 octahedron and two edges with LiO_6 octahedra.

Compared to layered $LiCoO_2$, $LiFePO_4$ has excellent structural stability and thermal stability owing to the strong covalent P-O bonds in the $(PO_4)^{3-}$ polyanionic clusters. The host framework can make host stable enough upon repeating cycling without evident structural changes and temperature limitations and no loss of oxygen. These advantages offer safe enough application for LIBs. However, long-term air exposure at a moderate temperature will give rise to surface oxidation of $LiFePO_4$/C nanocomposites with the formation of a $LiFePO_4(OH)$ phase. Despite these drawbacks, the advantages of LFP including cycle stability, safety, environmental friendliness, and low cost make this material a very promising cathode candidate in LIBs. Currently, it has been successfully commercialized by many battery manufactures such as A123 and BYD.

Although $LiFePO_4$ possesses many merits, it still has some drawbacks such as low intrinsic electric conductivity ($\sim 10^{-9}$ S cm^{-1}) and low packing density. During past decades, much effort has been focused on addressing these problems by optimizing synthesis method, surface modification, reducing particle size, and metal doping [22]. Among them, surface modification has been widely accepted as an effective approach for enhancing the electronic conductivity of $LiFePO_4$. Carbon coating is the most popular surface modification strategy for $LiFePO_4$. It can not only improve the electric conductivity of active particle but also act as a reducing agent to prevent the formation

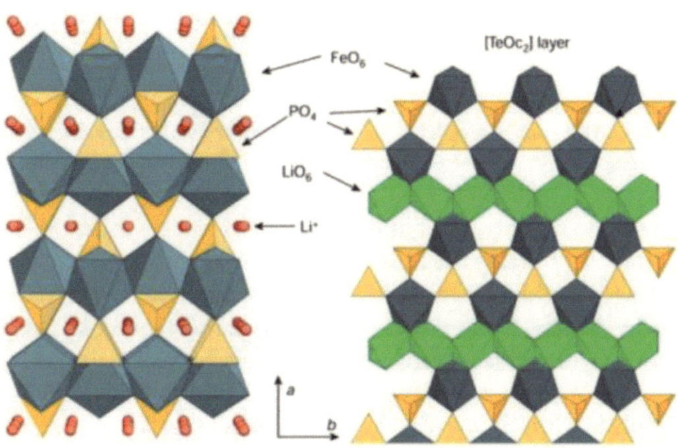

Figure 7.
The crystal structure of olivine LiFePO4 in projection along [001].

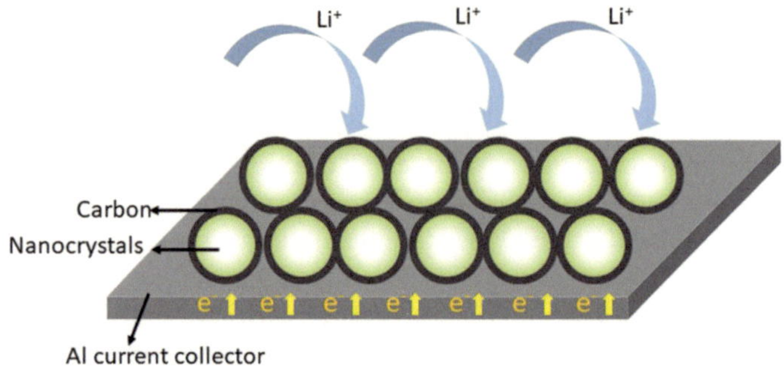

Figure 8.
Schematic diagram illustrating Li+ ion and electron transport pathways for active nanoparticle.

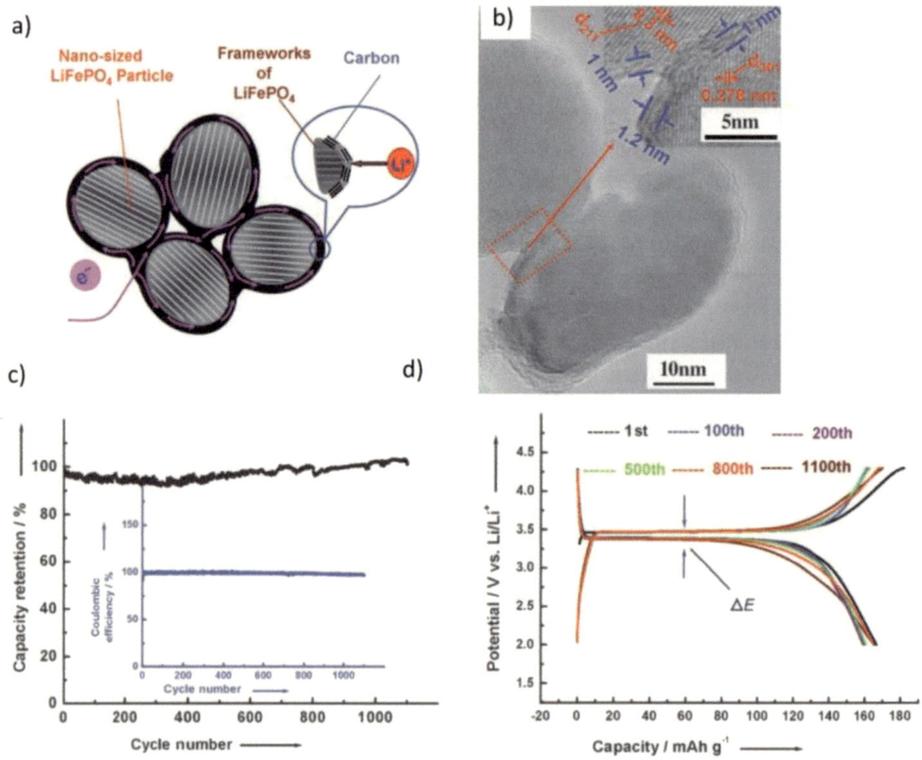

Figure 9.
(a) The ideal structure of carbon-coated LiFePO4 nanoparticles by a complete and uniform coating, (b) HRTEM image of carbon-coated LiFePO4 single nanoparticles, (c) cycling performance of carbon-coated LiFePO4 at a current density of 0.1 A g^{-1} between 2.0 and 4.3 V, and (d) charge/discharge profiles at different cycle numbers.

of Fe^{3+} impurity phase, as illustrated in **Figure 8**. In the meanwhile, the dense protection layer can effectively isolate the $LiFePO_4$ particle to inhibit the undesirable particle growth and self-aggregation. Nevertheless, a large amount of carbon reduces the volumetric energy density of the electrode. Various carbon sources have been used to generate carbon layer such as sucrose, starch, resorcinol-formaldehyde gel, sugar, carbon black, polypropylene, hydroxyethylcellulose, and so forth [22]. Therefore, carbon coating should be thin enough on the active particle surface. Wang et al. prepared uniform carbon-coated $LiFePO_4$ nanoparticle by an in situ polymerization restriction method for the synthesis of C/$LiFePO_4$ nanocomposite with a semi-graphitic carbon shell with a thickness of 1–2 nm [23], as shown in **Figure 9**. Charge/discharge tests show that the so-formed nanocomposites exhibit superior rate capability and cycling stability (**Figure 9**). In addition to the effect of carbon content in the nanocomposite, the resultant carbon by different organic precursors shows different electronic conductivities, indicating that sp^2-coordinated carbon exhibits higher electronic conductivity than sp^3-coordinated or amorphous carbonaceous materials. Besides carbon coating, other surface modifications including metallic conducting layer and conducting polymer film are also investigated to enhance the electrochemical performance of $LiFePO_4$. Besides $LiFePO_4$, other polyanion-based cathode materials have also been developed such as $Li_3V_2(PO_4)_3$, $LiFeSO_4F$, $LiFeSiO_4$, and so on.

5. Li-free cathode

Besides the above Li-containing cathode, there are some Li-free cathode materials that have been studied for LIBs. The most common one is vanadium oxide. Owing to the multiple valence characteristic of vanadium (+2, +3, +4, +5), vanadium pentoxide (V_2O_5) possesses versatile redox-dependent properties and exhibits a great potential in the rechargeable batteries. V_2O_5 has a typical layered structure, which is very beneficial to Li^+-ion intercalation/de-intercalation. In recent years, V_2O_5 has been widely studied as cathode materials for LIBs. V_2O_5 is a typical intercalation compound with a layer structure with an orthorhombic structure belonging to P_mmn space group. **Figure 10** shows the crystal structure of orthorhombic V_2O_5, which consists of edge- and corner-sharing layered square pyramids, forming V_2O_5 nanosheets linked together by a weak V-O interaction orthogonal to the c-direction. In a single-layer slab, there are three crystallographically distinct oxygen centers (O1, O2, and O3). The single coordinated terminal/apical vanadyl oxygen atom, O(1), has a relatively short V-O bond length of ca. 1.54 Å; the bridging O2 oxygen connects with two adjacent V atoms by corner-sharing VO_5 square pyramids. The corresponding V-O bond length is ca. 1.77 Å. The O3 with triply coordinated chaining O atoms connects three V atoms by edge-sharing VO_5 square pyramids, and the three corresponding V-O bond length are 1.88, 1.88, and 2.02 Å. The orthorhombic V_2O_5 can be described as highly distorted VO_6 octahedral building blocks to form the layered anisotropic structure. The long sixth V-O bond underlines the structural anisotropy of such material and the ability to insert guest species in perovskite-like cavities. During lithium insertion into layered V_2O_5, it will induce the formation of vanadium bronzes as follows:

$$V_2O_5 + xLi^+ + xe^- = Li_xV_2O_5 \qquad (1)$$

Besides crystallized V_2O_5, hydrated vanadium pentoxide ($V_2O_5\ nH_2O$) also shows promising lithium storage properties, such as $V_2O_5\ nH_2O$ xerogels, $V_2O_5\ nH_2O$ aerogels, and $V_2O_5\ nH_2O$ glasses [24]. These low crystalline or amorphous materials exhibit attractive merits owing to their unique nanostructures (high active surface

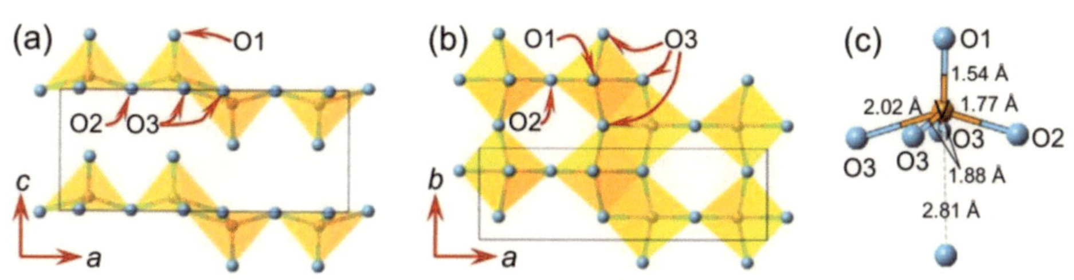

Figure 10.
Crystal structure of orthorhombic V2O5, view from (a) ac plane, (b) ab plane, and (c) the coordination environment around a single vanadium atom. The V and O atoms are represented in yellow and blue spheres.

area, small particle size, high diffusion coefficients, low-volume change upon repeated Li insertion/extraction processes). An effective approach to increase the intercalation capacity of V_2O_5 is to adjust the interlayer structure and the interaction forces between the adjacent layers. Usually, when the interlayer distance between V_2O_5 layered structure increases, the lithium insertion capacity will increase. However, owing to the absence of lithium in the V_2O_5, Li metal or Li-containing anode must be used for LIBs.

6. Thin-film battery applications of cathode materials

Among various battery chemistries, thin-film batteries are considered as the most competitive power sources owing to their high volumetric energy density, gravimetric specific energy, superior power capability, and good flexibility, as shown in **Figure 11**. Each cell is constructed by cathode, anode, and electrolyte, which is the same to conventional LIBs. To achieve high specific energy of thin-film battery, two requirements must be met by electrode materials: (i) high specific charge density (in Ah kg^{-1} and Ah L^{-1}) and (ii) high cell working voltage that requires high cathode redox potential and low anode redox potential. In addition, electrode reactions should be highly reversible on both cathode and anode for guaranteeing good cycling performance. Thus, developing advanced battery chemistry

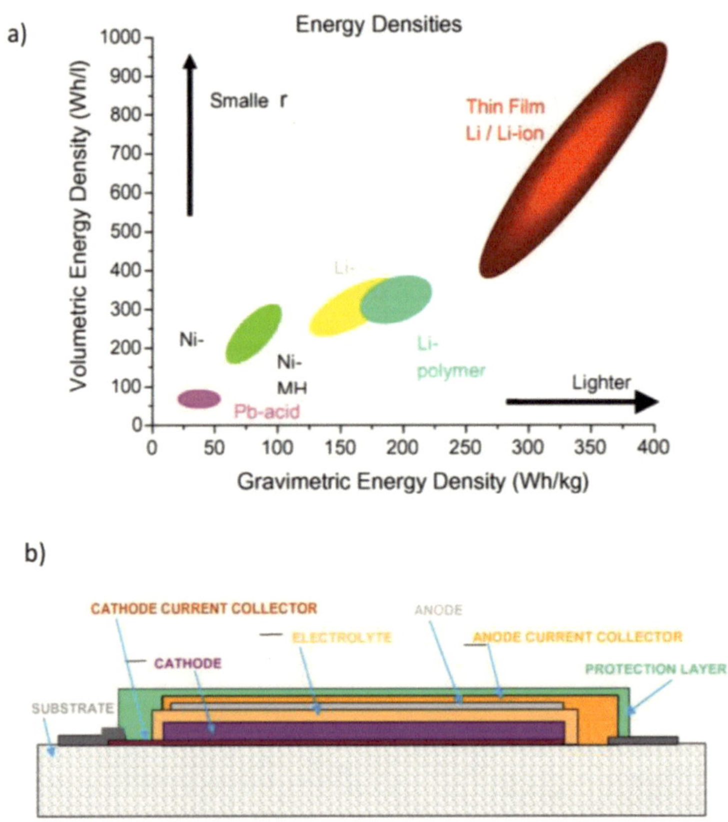

Figure 11.
(a) Comparison of the gravimetric and volumetric energy density of various chemical batteries including Pb-acid, Ni-MH, Li-ion batteries, and thin-film Li-ion batteries. (b) Cross-sectional schematic diagram of thin-film Li-ion battery.

is a key for high-performance thin-film batteries. This section only focuses on cathode materials for their thin-film microbattery application.

For practical cells, carbon-based anode and metal oxide cathode have been employed since they can offer maximum specific energy, sufficient power density, and long-term cycle life. Currently, several cathode materials ($LiCoO_2$, $LiNiO_2$, $LiMn_2O_4$, LNMO, and V_2O_5) are relatively popular. These electrodes are relatively stable in the air. In the case of V_2O_5, lithium metal laminate is attached to the carbon anode. After filling in the electrolyte, the metallic Li reacts with carbon and generates Li_xC_6, which is a lithium source in the cell.

$LiCoO_2$ is one of the most common used cathode materials for thin-film LIBs. Kuwata et al. prepared thin-film LIBs by using $LiCoO_2$, amorphous SnO, and amorphous Li_2O-V_2O_5-SiO_2 as cathode, anode, and solid electrolyte, respectively [25]. As illustrated in **Figure 12**, the battery has an area of about 0.23 cm^2, and the thickness is around 2 μm. All films were grown by PLD. The thin-film battery is fabricated on the quartz glass substrate, and metal Pt was deposited by PLD onto the substrate as a current collector. The solid electrolyte $Li_{3.4}V_{0.6}Si_{0.4}O_4$ was prepared by solid-state reaction from Li_2CO_3, SiO_2, and V_2O_5 and then was pressed into a disk with a diameter of 13 mm and annealed at 1100°C for the targets. The $LiCoO_2$ powder was pressed and sintered in air (900°C) as cathode, and the SnO powder was pressed into the disk without annealing. As displayed in **Figure 12**, the thicknesses of SnO, LVSO, and $LiCoO_2$ films are estimated to be 150, 1100, and 400 nm, respectively. The total thickness of the battery is around 2 μm. The interface between the $LiCoO_2$, SnO, and LVSO electrolyte remained smooth before and after cycling.

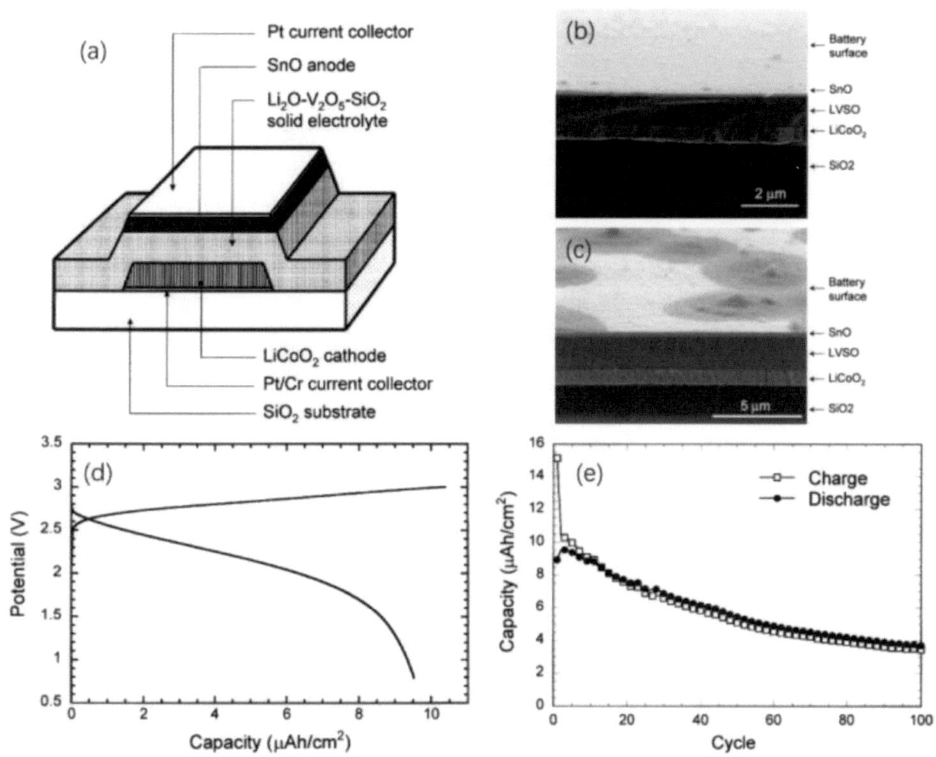

Figure 12.
(a) Schematic diagram (cross-sectional view) of thin-film LIBs, (b) SEM micrographs showing cross-sectional view before cycling, (c) after 100 cycles, (d) charge/discharge curves of the thin-film LIBs in the second cycle at a current of 44 mA/cm2, and (e) the specific capacity as a function of cycle number.

The thin-film battery was tested at a cutoff window of 3.0–0.7 V. As shown in **Figure 12**, the discharge capacity in the second cycle is about 9.5 mAh cm^{-2}, corresponding to 10–20% of $LiCoO_2$ utilization. The discharge potential gradually decreases in the range from 2.7 to 1.5 V owing to the amorphous nature of the SnO film. Such a thin-film battery shows a rapid capacity decay in the first cycle, which is probably attributed to the irreversible formation of Sn-Li alloy and amorphous Li_2O. Despite this, such battery was still cycled for 100 cycles with good Coulombic efficiency (**Figure 12**).

As discussed previously, various cathode materials ($LiCoO_2$, $LiMn_2O_4$, $LiFePO_4$, V_2O_5) have been studied for thin-film LIBs. However, they show limited specific energies. To further improve the energy density of the Li-ion thin-film micro-battery, an effective solution is using high-potential cathode material. It is well known that LNMO is a very promising high-voltage cathode material owing to its high working potential at 4.7 V vs. Li/Li^+. Nowadays, LNMO thin films have been deposited by various techniques such as sol-gel, electrostatic spray deposition, pulsed laser deposition (PLD), and RF magnetron sputtering deposition. Letiche et al. prepared LNMO thin film on functional current collector by a radiofrequency magnetron sputtering method [26]. The authors first solved the technological issue owing to PtSi phase which originates from the interdiffusion between silicon wafer and Cr/Pt current collector to form PtSi phase under annealing treatment. By doping the Cr layer with a dense and pinhole-free Al_2O_3 thin film deposited by ALD acting as a barrier diffusion between the Si substrate and the LNMO layer, the synthesis process has been validated. In addition, the authors demonstrated that the deposition pressure plays a key role on the Ni-Mn cation ordering in spinel-like structure ($P4_332$ ordered vs. $Fd3m$ disordered spinel) and consequently on the electrochemical performance. It indicates that LNMO thin film deposited at 10^{-2} mbar and annealed at 700°C exhibits a normalized capacity of 65 μAh cm^{-2} μm^{-1}

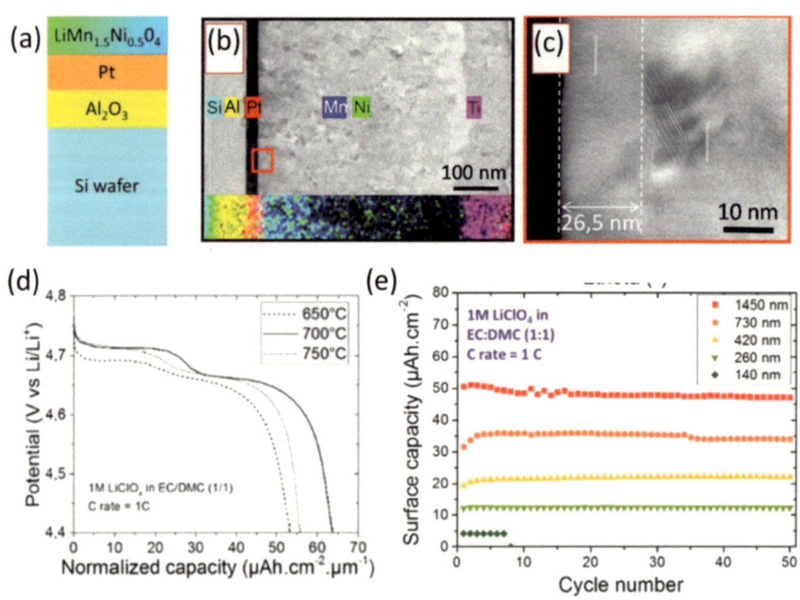

Figure 13.
(a) The as-prepared LNMO thin film, (b) TEM analysis of the cross section Si/Al2O3/Pt/LNMO/Ti/SiO2 prepared by focus ion beam technique, (c) HRTEM image of the LNMO layer with a focus close to the Pt current collector in order to check the Pt/LNMO interface, (d) discharge curve of LNMO thin film at different annealing temperatures, and (e) evolution of the surface capacity vs. cycle number as a function of the thin-film thickness.

with good capacity retention during cycling. The Coulombic efficiency is highly dependent on the film thickness of high-voltage thin-film electrodes.

As shown in **Figure 13**, the TEM image obtained by the focused ion beam (FIB) technique shows the different stacked layers on Si wafer where the thickness is close to 85 nm for Al_2O_3, 30 nm for Pt current collector, and 480 nm for LNMO thin-film electrode. Moreover, LNMO layer exhibits a polycrystalline nature with a different crystallized orientation (domains of ~ 20 nm), except for a thickness of 26.5 nm along the Pt layer where a (111) preferred orientation is seen, as displayed in **Figure 13**. The LNMO thin film annealed at 700°C delivers the best performance.

7. Conclusions

This chapter mainly discusses the development of currently the state-of-art cathode materials for LIBs and analyzes their potential applications in thin-film Li-ion microbatteries. For Li-containing cathode, layered structured $LiCoO_2$, NMC, and NCA, spinel $LiMn_2O_4$, and polyanion-based $LiFePO_4$ are currently the most popular for LIBs. Among them, $LiCoO_2$ is mainly used in the field of consumer electronics, while $LiMn_2O_4$, $LiFePO_4$, NMC, and NCA can be employed as power battery cathode materials for EVs. In particular, Ni-rich NCA and NMC-811 have been currently considered as the most attractive cathode candidate for long-range EVs due to their high specific energy compared to Ni-less NMC. In this respect, Ni-rich NMC cathode shows a sizable potential for high-energy thin-film LIBs in the future, but they are seldom investigated in thin-film Li-ion battery. Besides developing high-capacity cathode materials for realizing high-energy thin-film LIBs, exploiting high-potential cathode materials is also an effective strategy. Spinel LNMO is a very promising cathode candidate owing to its high redox potential (4.7 V vs. Li/Li^+), which has been widely investigated as cathode for thin-film LIBs. Additionally, Li-free cathode materials including V_2O_5, V_6O_{13}, MoS_2, TiS_2, and MnO_2 show attractive advantages for thin-film LIBs. Finally, the commercial success of thin-film Li-ion batteries will largely depend on the significant breakthrough of electrode materials including cathode, anode, and solid electrolyte in terms of energy density, weight, size, flexibility, and electrode/electrolyte interface.

Author details

Yuan-Li Ding
Hunan University, College of Materials Science and Engineering, Changsha, China

*Address all correspondence to: ylding@hnu.edu.cn

References

[1] Nishi Y. Lithium ion secondary batteries; past 10 years and the future. Journal of Power Sources. 2001;**100**(1-2):101-106

[2] Li H et al. Research on advanced materials for Li-ion batteries. Advanced Materials. 2009;**21**(45):4593-4607

[3] Liu C, Neale ZG, Cao G. Understanding electrochemical potentials of cathode materials in rechargeable batteries. Materials Today. 2016;**19**(2):109-123

[4] Goodenough JB, Park KS. The Li-ion rechargeable battery: A perspective. Journal of the American Chemical Society. 2013;**135**(4):1167-1176

[5] Chen Z, Dahn JR. Methods to obtain excellent capacity retention in $LiCoO_2$ cycled to 4.5 V. Electrochimica Acta. 2004;**49**(7):1079-1090

[6] Barker J, Koksbang R, Saidi MY. An electrochemical investigation into the lithium insertion properties of Li_xNiO_2 ($0 \leq x \leq 1$). Solid State Ionics. 1996;**89**:25-35

[7] Liu W et al. Nickel-rich layered lithium transition-metal oxide for high-energy lithium-ion batteries. Angewandte Chemie. 2015;**54**(15):4440-4457

[8] Manthiram A. An outlook on Lithium ion battery technology. ACS Central Science. 2017;**3**(10):1063-1069

[9] Peralta D et al. Submicronic $LiNi_{1/3}Mn_{1/3}Co_{1/3}O_2$ synthesized by co-precipitation for lithium ion batteries—Tailoring a classic process for enhanced energy and power density. Journal of Power Sources. 2018;**396**:527-532

[10] Gruber PW et al. Global Lithium availability: A constraint for electric vehicles. Journal of Industrial Ecology. 2011;**15**(5):760-775

[11] Manthiram A, Song B, Li W. A perspective on nickel-rich layered oxide cathodes for lithium-ion batteries. Energy Storage Materials. 2017;**6**:125-139

[12] Noh H et al. Comparison of the structural and electrochemical properties of layered $Li[Ni_xCo_yMn_z]O_2$ (x = 1/3, 0.5, 0.6, 0.7, 0.8 and 0.85) cathode material for lithium-ion batteries. Journal of Power Sources. 2013;**233**:121-130

[13] Sun Y et al. Synthesis and characterization of $Li[(Ni_{0.8}Co_{0.1}Mn_{0.1})_{0.8}(Ni_{0.5}Mn_{0.5})_{0.2}]O_2$ with the microscale Core–Shell structure as the positive electrode material for Lithium batteries. Journal of the American Chemical Society. 2005;**127**(38):13411-13418

[14] Sun Y et al. High-energy cathode material for long-life and safe lithium batteries. Nature Materials. 2009;**8**(4):320-324

[15] Thackeray MM. Spinel electrodes for Lithium batteries. Journal of the American Ceramic Society. 2004;**82**(12):3347-3354

[16] Kim J-H et al. Effect of Ti substitution for Mn on the structure of $LiNi_{0.5}Mn_{1.5-x}Ti_xO_4$ and their electrochemical properties as Lithium insertion material. Journal of the Electrochemical Society. 2004;**151**(11):A1911-A1918

[17] Alcantara R et al. Synergistic effects of double substitution in $LiNi_{0.5-y}Fe_yMn_{1.5}O_4$ spinel as 5 V cathode materials. Journal of the Electrochemical Society. 2005;**152**(1):A13-A18

[18] Xu X et al. $LiNi_{0.5}Mn_{1.5}O_{3.975}F_{0.05}$ as novel 5 V cathode material. Journal of Power Sources. 2007;**174**(2):1113-1116

[19] Sun Y-K et al. Electrochemical performance of nano-sized ZnO-coated $LiNi_{0.5}Mn_{1.5}O_4$ spinel as 5 V materials at elevated temperatures. Electrochemistry Communications. 2002;**4**(4):344-348

[20] Sun Y-K, Yoon C, Oh I-H. Surface structural change of ZnO-coated $LiNi_{0.5}Mn_{1.5}O_4$ spinel as 5 V cathode materials at elevated temperatures. Electrochimica Acta. 2003;**48**(5):503-506

[21] Kobayashi Y et al. 5 V class all-solid-state composite lithium battery with Li_3PO_4 coated $LiNi_{0.5}Mn_{1.5}O_4$. Journal of the Electrochemical Society. 2003;**150**(12):A1577-A1582

[22] Wang Y, He P, Zhou H. Olivine $LiFePO_4$: Development and future. Energy & Environmental Science. 2011;**4**(3):805-817

[23] Wang Y et al. The design of a $LiFePO_4$/carbon nanocomposite with a Core–Shell structure and its synthesis by an in situ polymerization restriction method. Angewandte Chemie. 2008;**47**(39):7461-7465

[24] Wang Y et al. Nanostructured vanadium oxide electrodes for enhanced Lithium-ion intercalation. Advanced Functional Materials. 2006;**16**(9):1133-1144

[25] Kuwata N et al. Thin-film lithium-ion battery with amorphous solid electrolyte fabricated by pulsed laser deposition. Electrochemistry Communications. 2004;**6**(4):417-421

[26] Letiche M et al. Tuning the cation ordering with the deposition pressure in sputtered $LiMn_{1.5}Ni_{0.5}O_4$ thin film deposited on functional current collectors for Li-ion microbattery applications. Chemistry of Materials. 2017;**29**(14):6044-6057

Section 4

Thin Film for Anode

Chapter 5

Flexible Porous Carbon Nanotube Films Intercalated with Active and Functional Materials for Lithium-Ion Batteries

Xiaogang Sun, Xu Li, Wei Chen, Jie Wang, Chengcheng Wei, Yapan Huang, Guodong Liang and Hao Hu

Abstract

Lithium-ion battery (LIB) has occupied the main position of portable electronic devices, and it is also playing an important role in energy storage and large energy storage. Thin film devices based on their diverse functions have great potential for wide application. Novel thin film devices need to be created for the improvement of electrochemical performance and safety of LIB. Our research focused on transparent conductive films and new flexible porous carbon nanotube films for improving and enhancing the energy/power density and cyclic performance of LIB. Meanwhile, different carbon nanotube films have their own additional advantages in strength and thermal conductivity to meet various functional requirements of LIBS.

Keywords: carbon nanotubes, lithium-ion battery, cellulose, film, collector

1. Introduction

At present, fossil energy has been overexploited and overutilized. Traditional fuels are not only facing depletion of reserves but also causing serious air pollution and endangering human health. Comfortable, fresh, green, and pollution-free living environment has become the urgent need of humans [1–5]. Now energy problem and environmental protection have become two important issues in the world. With the rapid development of portable electronic equipment and transportation tools, many countries have devoted a lot of human and material resources to the development of high-performance energy storage devices [6–12] with low consumption and that are pollution-free.

Thin film devices have so many functions which can be conductive, magnetic, luminous, catalytic, antibacterial, sound insulated, and flame retardant and other characteristics after compounding with different functional materials that are widely used in antistatic packaging materials, electromagnetic shielding materials, new energy and electrochemical materials, heating materials, and sensing and braking materials [13–28] and other fields. For example, transparent conductive film has been extensively used in flat panel displays, solar cells, and touch panels and other fields because of its conductive and light transmission characteristics [29–31]. However, commercial thin film devices which are widely used nowadays have high cost and poor bending resistance, so it is necessary to create new thin film devices.

It has been 20 years after the discovery of carbon nanotubes (CNTs) by Japanese scientist Iijima in 1991 [32]. CNTs have become a hot carbon material for researchers to explore and develop its wider range of properties and applications actively. It is found that CNTs are one-dimensional materials with seamless, hollow, and single-walled or multi-walled tubular structures and curled by graphite [33–40]. CNTs have excellent electrical and mechanical properties, high aspect ratio, and good lithium insertion performance. Carbon nanomaterial-based transparent conductive films have good electrical conductivity, outstanding chemical stability [27, 29, 41–44], good substrate adhesion, and excellent mechanical flexibility and can be produced in large quantities and suitable for continuous filmmaking, which make CNTs occupy an important position in the field of new thin film devices, especially in the field of flexible transparent conductive films. And CNTs were also the earliest carbon nanomaterials [30, 37, 45–48] used to study transparent conductive films. Nowadays, the commercial production of CNTs has been very perfect. But CNTs are nanoscale materials, with large specific surface area and high surface energy, so CNTs show significant agglomeration effect. In addition, CNTs are one-dimensional fibrous nanomaterials with large length/diameter ratio, which makes CNTs have intertwined and binding properties similar to that of fibers. The interaction of the two properties makes CNTs form stable aggregates, affecting their further commercialized application [47, 49]. In order to solve this problem, researchers have searched for many methods of decentralization. Physical dispersion methods include grinding, ball milling, ultrasonic oscillation, high-speed shearing, and so on [37, 38, 48–50]. Chemical dispersion methods include washing with strong acid and strong base and adding surfactant dispersant [51–54]. However, all the dispersion methods have their limitations, so it is difficult to obtain a stable carbon nanotube dispersion solution. Therefore, it is necessary to develop better dispersion methods or to make a kind of easily dispersed carbon nanotubes for better application in the development of thin film devices.

In this work, we have created a method for manufacturing carbon nanotubes which can be dispersed easily (WhiskerCNT). Carbon nanotube transparent conductive film (WTF), flexible porous carbon nanotubes films (FWFs), and active conductive film fabricated by WhiskerCNT have good physical and chemical properties.

2. Preparation and characterization of WhiskerCNT

2.1 Fabrication of WhiskerCNT

Benzene was used as carbon source. Ferrocene was used as catalyst, and thiophene was added into benzene to stir evenly. The input rate of raw materials is controlled by a micro pump. The carrier gas was composed of hydrogen and argon, and the gas flow velocity was controlled by a mass flowmeter. The reaction was carried out in a vertical furnace, and the preparation was carried out at a certain temperature gradient (the detailed production process is described in the corresponding patent of our research group). The purity of WhiskerCNT obtained by preliminary production is 95–97%, and further graphitization is needed to improve the purity.

WhiskerCNT was placed in a graphitizing furnace, and high purity argon was introduced as a protective gas after vacuum extraction. When the furnace is cooled, WhiskerCNT can be removed. Then the purity can be over 99.6%.

2.2 Characterization of WhiskerCNT

Figure 1 is the morphological characterization of graphitized WhiskerCNT. WhiskerCNT is linear tubular structures with relatively small transverse lengths and large aspect ratios, and it is mainly distributed in a straight line, which makes the chances of bending and winding between WhiskerCNT and pipes small and easy to disperse (**Figure 1a**). WhiskerCNT diameter is about 50 nm. The nanotubes have a thin cavity in the middle. The layered structure of the tube wall is clear and orderly (**Figure 1b**), and the interval is 0.338 nm by calculating wall thickness and stories.

Figure 2 is the characterization of graphitized WhiskerCNT. The sharp diffraction peaks of graphitized WhiskerCNT show that WhiskerCNTs have high crystallinity (**Figure 2a**). The D peak of Raman pattern is very low, indicating that WhiskerCNTs have a complete structure and an orderly arrangement (**Figure 2b**). The G peak is sharp, and the I_G/I_D is up to 16.95, which proves that WhiskerCNT has high graphitization degree and high purity.

Figure 1.
Morphological characterization of graphitized WhiskerCNT: (a) SEM and (b) TEM.

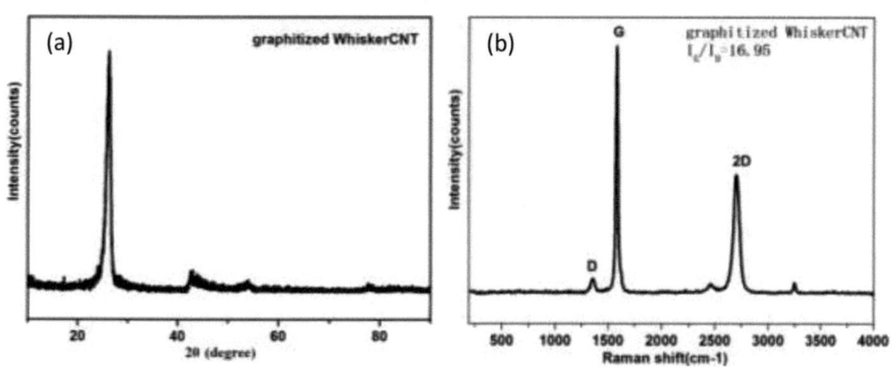

Figure 2.
XRD pattern (a) and Raman pattern (b) of graphitized WhiskerCNT.

3. WhiskerCNT transparent conductive films (WTF)

3.1 Preparation of WTF

The WhiskerCNT and graphitized WhiskerCNT were milled 2 h in a ball mill (200 r/min) to obtain fine powders. Two kinds of conductive fluids were prepared by adding the ball-milled WhiskerCNT (1 wt%) to ethanol with TNADIS (0.05 wt%)

and ultrasonic treating for 30 min and high-speed shearing for 1 h. The two conductive fluids were coated on the transparent PET film by spin coating. After spin coating for 1, 2, and 3 min separately, they were put into a vacuum drying box to dry.

	Item	1 min	2 min	3 min
WhiskerCNT	Square resistance (kΩ/sq)	103.3	10.6	3.7
	Transmittance (%)	68.30	57.90	52.80
G-WhiskerCNT	Square resistance (kΩ/sq)	53.6	2.8	0.34
	Transmittance (%)	68.90	58.10	53.30

Table 1.
Square resistance and transmittance of WTF.

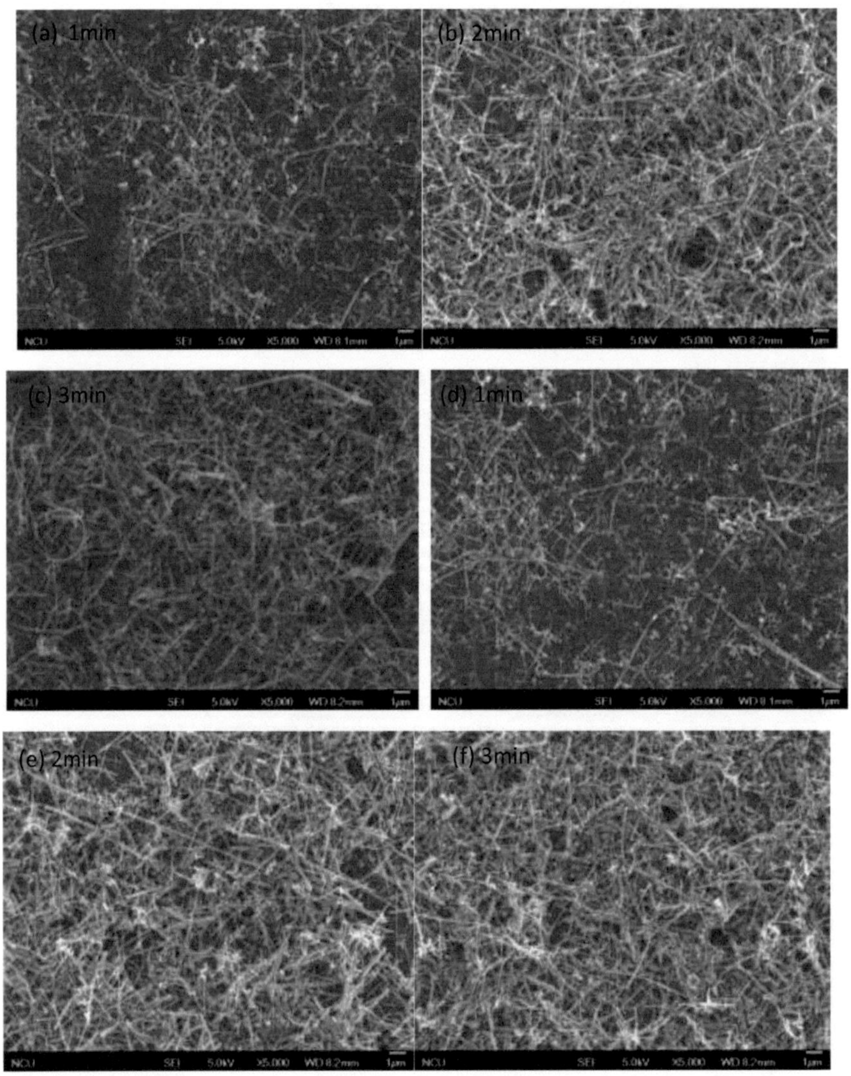

Figure 3.
SEM pattern of WTF. WhiskerCNT (a)–(c), graphitization WhiskerCNT (d)–(f).

3.2 Results and discussion

It is found from **Table 1** that the transmittance of the two films is similar, while the conductivity of G-WhiskerCNT WTF is stronger under the same spin coating time. With the increase of spin coating time, the transmittance of both films decreased, because WhiskerCNT gradually formed a continuous and dense network of electronic transmission paths from discontinuous interconnection (**Figure 3**).

4. Flexible porous WhiskerCNT films

In order to improve the electrochemical performance of lithium-ion batteries, a kind of flexible porous carbon nanotube films (FWFs) compounded of WhiskerCNT and cellulose fiber was prepared by process similar to papermaking method.

4.1 Preparation of FWF

Graphitized MWCNT powder was dispersed in deionized water by sonication for 2 h and followed by high-speed shearing for 2 h with an addition of sodium dodecylsulfate (SDS) as a surfactant. Cellulose pulp was prepared by immersing softwood pulp into deionized water and followed by high-speed shearing for 3 h. The MWCNT dispersion liquid and the cellulose pulp were mixed by high-shear emulsifier to form suspension for 3 h. The suspension liquid of cellulose and MWCNT pulp was infiltrated by vacuum filtration. A randomly interwoven fiber mat was obtained. The host was rolled and tailored for the activated materials.

4.2 Characterization of FWF

Figure 4(a) showed the SEM image of top surface of FWF with a random-in-plane weblike network structure. FWF host demonstrated homogenous incorporation of WhiskerCNT in the cellulose fiber network. **Figure 4(c)** showed the cross-sectional image of FWF. The figures clearly displayed porous network structure which provided sufficient void space for loading a large amount of active substance and accommodating the volume expansion of active substance during cycles. The coarse surface and interconnected channel permit good penetration of the electrolyte and active substance. WhiskerCNT is uniformly dispersed and attached on paper fiber (**Figure 4b**). A good three-dimensional conductive network was constructed with cellulose fiber as the framework and MWCNT as conductors. WhiskerCNTs have smaller nanostructures, so they fill, adsorb, and agglomerate on the surface of paper fibers and in the gap between paper fibers, acting as the carrier of paper fiber connection. FWF exhibited porous, flexible, and high specific surface area. Pore size distribution of FWF is shown in **Figure 4(d)**. From the picture, we can see the host has big porosity. The special surface area of FWF reached 25.6 m^2/g.

4.3 FWF as collector in ternary materials of lithium-ion battery

4.3.1 Preparation of positive electrode and half-cells

After milling WhiskerCNT, SDS (CNTs:SDS = 90 wt%:10 wt%) and NMP solvents were added and followed by high-speed shearing for 4 hours. The WhiskerCNTs were dispersed fully and prepared into 5% solid content WhiskerCNT

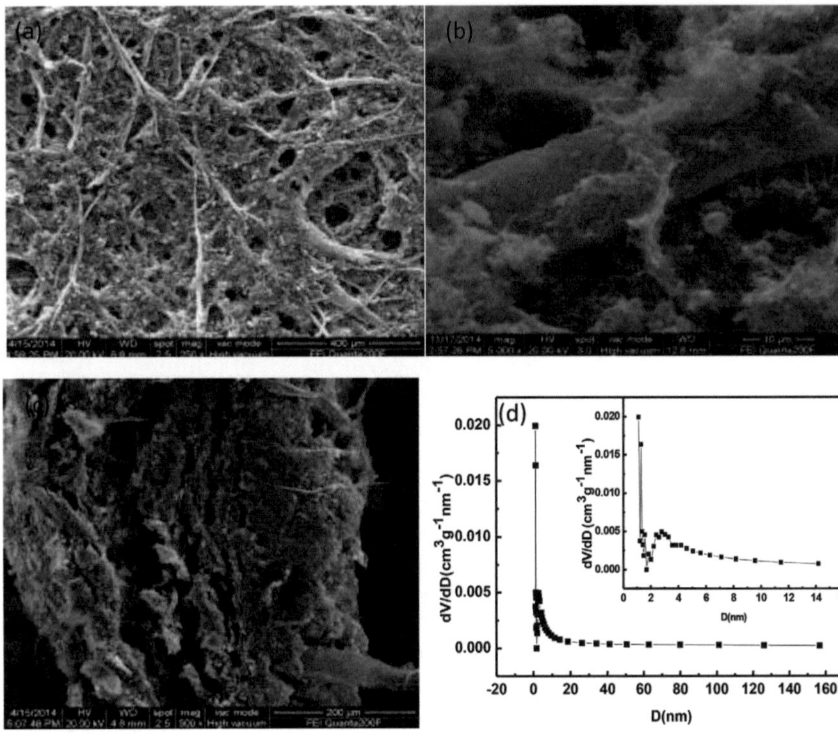

Figure 4.
SEM of FWF (a), (b). TEM of FWF (c). BET of FWF (d).

dispersion solution. In the proportion of LNCM:WhiskerCNT:PVDF = 90 wt%:5 wt %:5 wt%:5 wt%, ternary cathode active materials were prepared, in which WhiskerCNT came from the preceding configuration of carbon nanotube dispersion. The cathode slurry was prepared by placing the prepared material in a sealed test tube, adding an appropriate amount of NMP, ultrasonic treating for 0.5 h, and shearing for 0.5 h at a high speed of 10,000 r/min. The cathode paste was divided into two equal parts, coated on aluminum foil and FWF, and dried at 60C. In the vacuum glove box, two kinds of electrode plates were assembled into a button cell, and the electrochemical performance of the cell was tested.

4.3.2 Results and discussion

Aluminum foil and FWF were coated with LNCM paste to assemble the battery and test the electrochemical performance of the battery. **Figure 5** is a comparison of the first charge of lithium-ion batteries with two different collectors at a discharge voltage of 2.0–4.5 V at 0.1C. Among them, the first discharge-specific capacity of ternary/FWF electrode (a) is 184 mAh/g, and the first discharge-specific capacity of ternary/aluminum foil electrode (b) is 178 mAh/g. (a) relative to (b) the specific capacity increased by 3.3%. It shows that FWF instead of aluminum foil is the collector and has a certain specific capacity improvement effect at low magnification.

The SEM of **Figure 6** shows the microcosmic situation of three yuan doped with 5% CNTs. It is observed that the ternary spherical particles are completely coated by Whisker CNT and even stacked in some places. WhiskerCNTs are one-dimensional carbon nanomaterials with large aspect ratio, as shown in **Figure 6**. As a conductive agent, a three-dimensional space conductive network system is formed by interpolating and overlapping the carbon tube with the carbon tube.

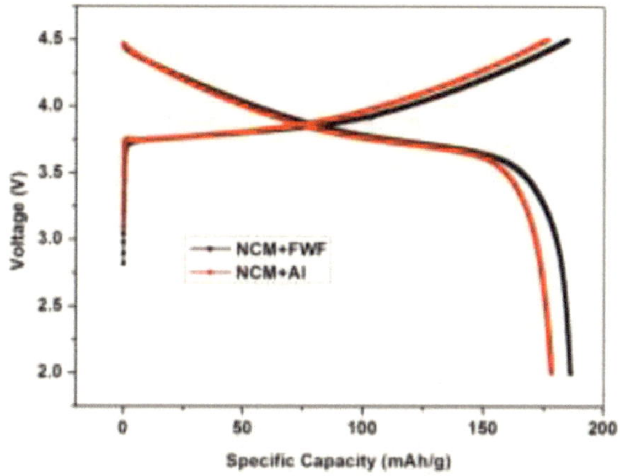

Figure 5.
First charge/discharge curves of two kinds of different collector electrodes.

Figure 6.
SEM images of ternary cathode material doped with 5 wt% CNTs.

At the same time, the three-dimensional network is distributed among the ternary particles, conducting the electronic transmission between the ternary particles and the particles. Compared with the traditional super carbon black (SP), the way of conducting carbon nanotubes is extremely efficient. At the same time, the carbon nanotubes were duplicated in the experiment, the utilization rate was not maximized, and the content of carbon nanotubes could be reduced to reduce the cost. At the same time, it was found that the ternary cathode material on the ternary/aluminum foil electrode was easy to detach, while the ternary/FWF electrode was tightly bonded and difficult to separate. It shows that three yuan/FWF is more solid than three yuan/aluminum foils and the two composite effects are better. The reason is that the surface of aluminum foil is smooth, and the internal pores of conductive paper are abundant, and the adsorption is strong, so the ternary material after coating penetrates into the interior of conductive paper and bonds closely with conductive paper, while the aluminum foil can only be attached to the surface.

Figure 7 is the micrograph of ternary/aluminum foil electrode and ternary/FWF electrode. The reason why FWF is better than aluminum foil is analyzed. It can be seen from the observation (a) that, because the aluminum foil is smooth and flat, the ternary coating on the aluminum foil contacts in the form of point to face, (a) fully showing that there is a large gap between the two interfaces due to this factor. In **Figure 7(b)** FWF, because of its uneven internal structure and rich pores, ternary materials penetrate into the interior of conductive paper, and ternary materials and conductive paper are closely combined. Therefore, the interfacial gap between ternary and FWF is basically absent in **Figure 7(b)**, showing a good interfacial bonding effect. Because of the advantages of interface performance, the three-element/FWF electrode has better electrochemical performance at low magnification.

The first 28 cycles of the 2 electrodes showed that the capacity retention rate of ternary/FWF electrode was 94% and that of ternary/aluminum foil electrode was 94% at 0.1C. But the curve in **Figure 8** shows that the specific capacity of the ternary/conductive paper electrode is higher than that of the ternary/aluminum foil electrode at a small rate, indicating that the specific capacity of the battery has been improved when FWF is used as a collector (the same information is shown in **Figure 5**).

The ratio performance diagram of ternary/FWF electrode is shown on the left and the ratio performance diagram of ternary/aluminum foil electrode on the right. At 0.1, 0.5, 1, 2, and 0.1C discharge, the specific capacities of ternary/FWF electrodes are 184, 155, 120, 120, and 160 mAh/g, and the specific capacities of ternary/aluminum foil electrodes are 178, 154, 138, 121, and 165 mAh/g. The specific capacity of ternary/FWF electrode was 184 mAh/g at 0.1C low rate, while that of

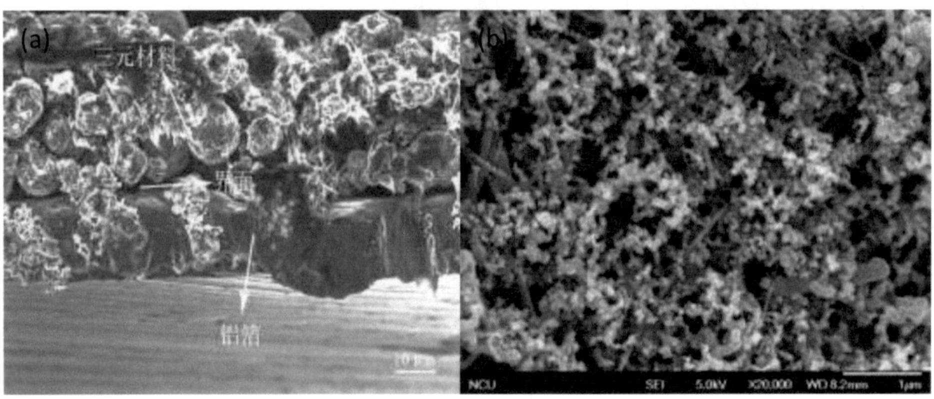

Figure 7.
(a) SEM image of NCM/aluminum foil electrode and (b) SEM image of NCM/FWF.

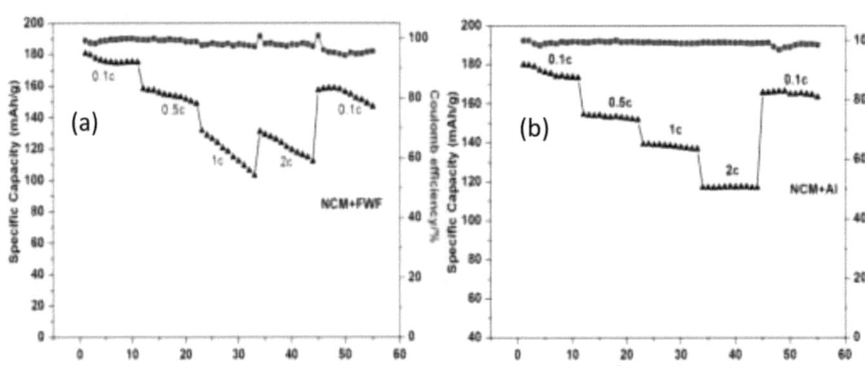

Figure 8.
Comparison of the two kinds of electrode's rate performance. NCM + FWF (a) and NCM + Al (b).

ternary/aluminum foil electrode was 178 mAh/g. However, at 1C and 2C high rate, the specific capacities of the ternary/conductive paper electrodes decreased to about 120 and 120 mAh/g, and the specific capacities of the ternary/aluminum foil electrodes remained good and tended to be stable at about 138 and 121 mAh/g. It shows that ternary/FWF is better than ternary/aluminum foil electrode at low magnification, but there are some shortcomings at high magnification. The reason is that FWF becomes looser and decomposed after a long time of electrolyte infiltration in the late period of each rate discharge, resulting in the increase of resistance of electronic transmission channel and the decrease of capacity. During the discharge of 1C and 2C, it was suspended for 1 minute. Because of the looseness and micro-elasticity of the FWF, the specific capacity of the battery was temporarily increased and then began to decay after a short recovery period of 1 minute, so the discharge of 2C reached 1C mode again, showing the phenomenon that the specific capacity of 2C was higher than that of 1C in the later period.

Figure 9 shows the impedance spectrum comparison of the two electrodes. The starting point of high-frequency region and the intersection point of real axis are the resistance in solution (Rs), the semicircle in intermediate frequency region is the charge transfer resistance (Rct) between SEI film and solid electrode, and the line in low frequency region is the diffusion resistance Zw of lithium ion in solid electrode material. It can be seen from observation that the Rs of the two batteries is basically the same. The Rct of ternary/FWF electrode is 175 and that of ternary/aluminum foil electrode is 275, indicating that FWF instead of aluminum foil as collector has a smaller charge transfer resistance, which is conducive to improving the electron migration rate and reaction depth during charging and discharging.

4.3.3 Conclusions

Graphitization treatment can improve the defects of carbon nanotubes, increase the degree of graphitization of carbon nanotubes, and make the arrangement of carbon atoms more orderly, which are conducive to the follow-up application. The initial specific capacity of ternary/FWF electrode is 184 mAh/g and that of ternary/aluminum foil electrode is 178 mAh/g at 0.1C, which is 3% higher than that of aluminum foil electrode. After 28 cycles, the capacity retention rate is above 94%.

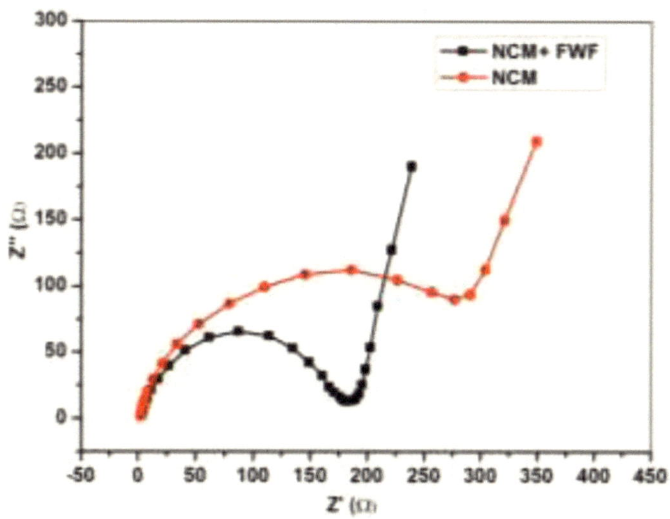

Figure 9.
Impedance spectra of NCM/conductive paper electrode and NCM/aluminum foil electrode.

At 1C and 2C high rate discharge, the specific capacities of ternary/FWF electrodes are 120 and 120 mAh/g and that of ternary/aluminum foil electrodes are 138 and 121 mAh/g. In contrast, the ternary/FWF electrodes have a greater attenuation because of the decomposition and shedding of FWF at the later stage, which needs to be solved. When FWF replaces aluminum foil as collector, the ternary/aluminum foil electrode Rct is 275, and the ternary/FWF electrode Rct is 175, which decreases by 36%, indicating that the impedance of the battery has been significantly improved. FWF instead of aluminum foil as collector has better interfacial performance; it makes ternary materials and collector substrate bond more closely, reduces the internal impedance of the battery, increases the specific capacity of ternary materials, and maintains better cycle performance, but the high rate performance needs to be improved.

4.4 FWF as collector in Li-Si battery

4.4.1 Preparation of electrode and half-cells

WhiskerCNT powders and hollow silicon were mixed and then were dispersed by sonication and high-speed cutting in distilled water for 2 h, respectively. The dispersant of polyvinylpyrrolidone (PVP) was added to improve the dispersion performance. The slurry of hollow silicon and WhiskerCNT was coated on the FWF.

CR2025 coin-type cells were assembled in an Ar-filled glove box with a metallic lithium foil as the counter electrode. The electrochemical performance was tested by a cell tester with 1 M $LiPF_6$ in a mixture of ethylene carbonate (EC) and diethyl carbonate (DEC) as electrolyte.

4.4.2 Results and discussions

In **Figure 10a**, it can be observed that the distribution of hollow silicon is chaotic and disordered, and get agglomeration phenomenon. While there is a different situation in **Figure 10b**, hollow Si can embed into micropores of FWF evenly. Micron-sized hollow silicon is evenly dispersed in the high-porosity grid consisted by WhiskerCNT and fiber, forming a similar coating effect (**Figure 10c**), which solves the problem of nonconductivity of Si and increases the storage channel of Li+. The active material can be well penetrated into the layered porous structure of FWF. At the same time, the WhiskerCNTs in active materials and in FWF are well connected, which makes the interface resistance to decrease, the electron transfer channel to increase, and the cell polarization effect to decrease, thus increasing the utilization rate of active materials. Si hollow microspheres are wrapped in layers by WhiskerCNT (**Figure 10d**), which can accommodate and buffer the volume expansion of Si. Moreover, FWF adsorbs the active material well so that active material is hard to crack or be separated from the current collector.

As shown in **Figure 11**, the first battery constant current discharge of 0.01 V, observed in the first discharge process to the long platform that disappeared in the subsequent cycle, corresponding to the SEI film, is achieved by irreversible reaction, which leads to the low efficiency of Coulomb in the loss rate and charging capacity. The copper collector silicon lithium-ion battery (CLIB) at a current density of 0.02C, has an initial discharge capacity of 869 mAh/g, and Kulun for the first time efficiency is 38%. The second discharge reduced to 364 mAh/g, from the beginning of the battery capacity little attenuation, Kulun time efficiency was between 86 and 89%. The more stable cycle performance is attributed to the large volume change of the hollow silicon, which does not break, and the rapid diffusion of lithium ion in the hollow structure silicon. Even so, the volume change of silicon

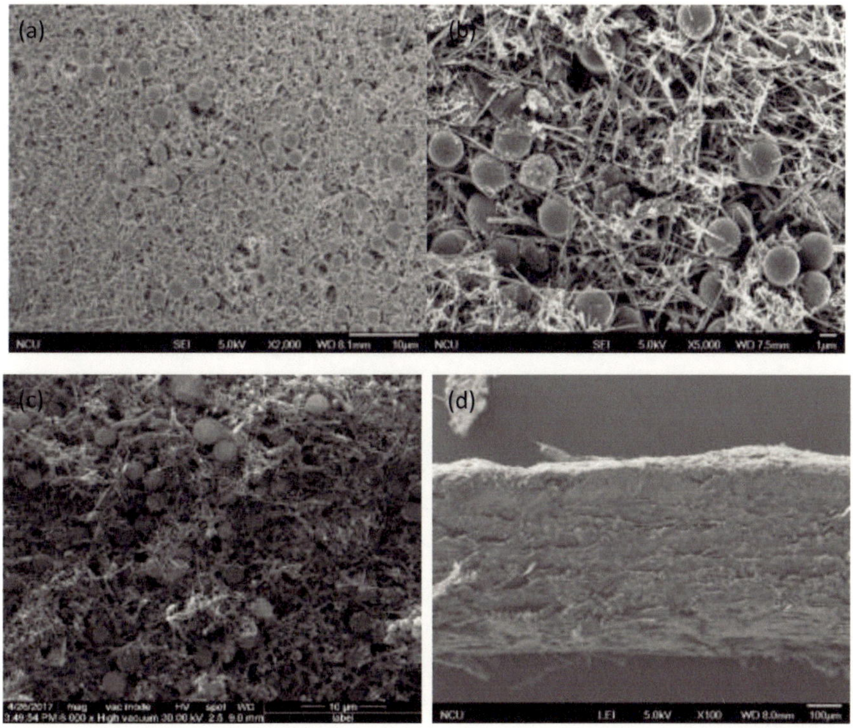

Figure 10.
SEM image of hollow Si-doped MWCNT compound material coated on copper foil (a), conductive paper (b, c), and its sectional drawing (d).

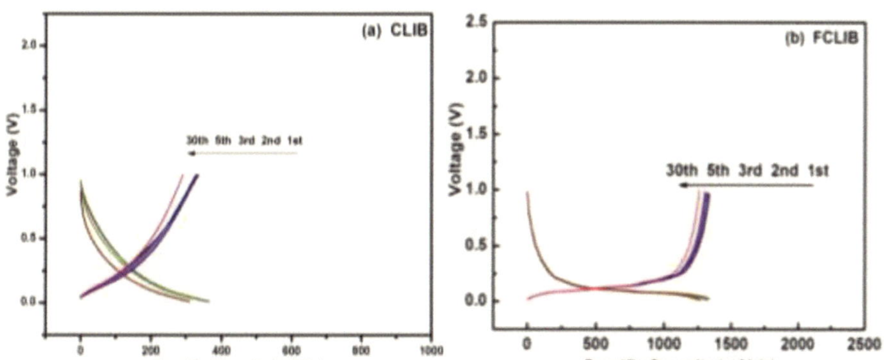

Figure 11.
Charging-discharging curve of CLIB (a) and FCLIB (b).

in the process of intercalation and debinding will lead to mutual extrusion and detachment between silicon particles. After repeated expansion and contraction, some silicon will even fall off from the copper foil collector. FWF collector for silicon-based lithium-ion battery (FCLIB) has an initial discharge capacity of 2168 mAh/g, and Kulun for the first time efficiency is 61.5%. The second discharge capacity is 1335 mAh/g, also from the second charge discharge start battery rapidly stable than capacity little attenuation, Kulun time efficiency are close to 100%, and cycle the 30 time was 1300 mAh/g, showed excellent discharge performance (**Figure 12**).

Under the current of 0.25C, PCLIB decreased continuously in the first eight times and then slowly reached 1150 mAh/g. When the small current density cycle is

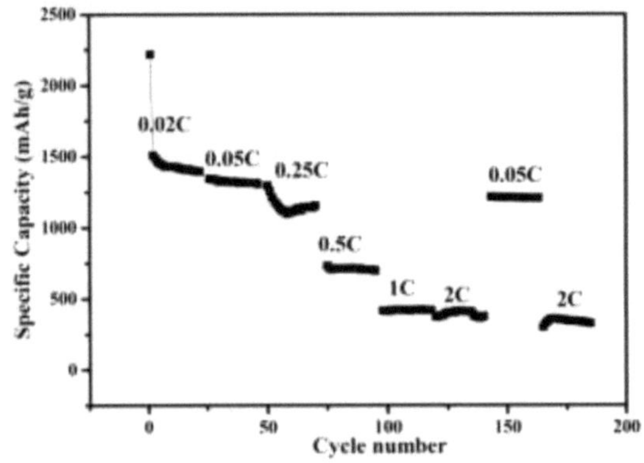

Figure 12.
Cyclic performance diagram of conductive paper battery under different current densities.

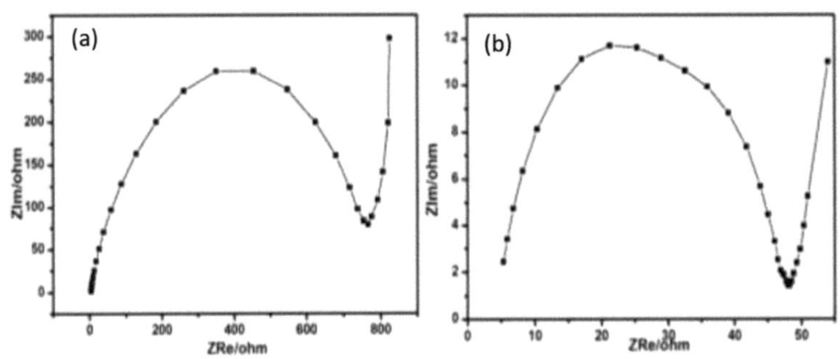

Figure 13.
EIS test diagram of conductive paper battery: predischarge (a) and after 30 cycles (b).

recovered, the specific capacity of CPLIB can be recovered to the original level. As figure PCLIB first circulated at the 2C current density for 30 times, it recovered to the 0.25C current, and the specific capacity recovered. This data shows the ability of PCLIB to maintain good specific capacity after repeated high current impact, long service life, and excellent cycling performance (**Figure 13**).

To better understand the volume change and the stability of SEI layer, the impedance analysis was performed. As shown in the picture, the contact resistance of the PSiLIB interface decreases significantly after the cycle. This is because the contact interface between active material and the collector will form a contact interface, which will cause greater resistance. As the reaction proceeds, the material will penetrate or compound to each other, and the contact surface will decrease or form the transition layer. The interaction of FWF with the WhiskerCNT in the active material enhances the conductivity of the active material and makes the resistance of the electrode smaller. In addition, the layered porous structure of CNT has the ability to adsorb and store the electrolyte stronger than the copper foil, which can make the reaction more complete and thorough.

4.4.3 Conclusions

In the summary, FWF was used as host of hollow Si for high-performance Li-ion battery. FWF has a porous structure and interconnected channel and exhibited high conduction and high absorption of electrolyte. Hollow Si penetrated into the network of FWF and achieved a low interface resistance which contributed to the high rate performance of the battery. FWF with interconnected channel can absorb an amount of electrolyte and accommodate the volume expansion of hollow silicon to prevent the cracking and pulverization of hollow silicon in cycles. All those help to enhance electrochemical performance of FWF electrode. The good cycle performance was ascribed to the accommodation and buffering effect of FWF to huge volume expansion of silicon during cycles. Therefore, it was believed that FWF has a useful application prospect as current collector for high-performance silicon-based lithium-ion batteries.

4.5 Outlook for further improvement of FWF

In order to further improve the functionality and practicability of FWF, we used aramid fibers instead of paper fibers to make flexible porous WhiskerCNT films with ultrahigh strength and toughness (SFWF), polyimide fiber to make flexible porous WhiskerCNT films with high thermal conductivity (TFWF), and hydroxyapatite nanowires to make environmentally friendly multifunctional flexible porous WhiskerCNT films (EFWF).

Aramid fiber has excellent properties such as ultrahigh strength, high modulus, and high temperature resistance, acid and alkali resistance, and aging resistance, and insulation and has lightweight property, long life cycle, and so on. Its strength is 5–6 times that of steel wires, its toughness is 2 times that of steel wire, and its weight is only about 1/5 of that of steel wire. So it has been applied to many fields such as bulletproof products, special protective clothing, and so on. While SFWF produced by aramid fiber also has high strength, high toughness, and other properties by different kinds of performance testing. TFWF has higher thermal conductivity than FWF. Similarly, EFWF hydroxyapatite nanowires have high temperature resistance, corrosion resistance, and fire resistance and are an environmentally friendly and biocompatible material, because hydroxyapatite is the main inorganic component of vertebrate bones and teeth. Therefore, EFWF has wider application fields. The above three thin film devices have completed the preliminary performance testing, and further electrochemical performance testing needs to be carried out.

5. Conclusion and outlook

In summary, we reported our study of transparent conductive thin films and FWF for lithium-ion battery applications. It has great potential application value for lithium-ion battery, energy storage equipment, and other fields. Meanwhile, TFWF and EFWF have better physical and chemical properties than FWF. And we believe they have better application prospects.

Author details

Xiaogang Sun*, Xu Li, Wei Chen, Jie Wang, Chengcheng Wei, Yapan Huang, Guodong Liang and Hao Hu
School of Mechatronics Engineering, Nanchang University, Nanchang, China

*Address all correspondence to: xiaogangsun@163.com

References

[1] Park S, Vosguerichian M, Bao Z. A review of fabrication and applications of carbon nanotube film-based flexible electronics. Nanoscale. 2013;**5**(5):1727

[2] Smith K, Wang CY. Power and thermal characterization of a lithium-ion battery pack for hybrid-electric vehicles. Journal of Power Sources. 2006;**160**(1):662-673

[3] Kennedy B, Patterson D, Camilleri S. Use of lithium-ion batteries in electric vehicles. Journal of Power Sources. 2000;**90**(2):156-162

[4] Zhong S, Hu J, Wu Z, et al. Performance of lithium ion batteries using a carbon nanotube film as a cathode current collector. Carbon. 2015; **81**:852

[5] Tang Z, He Y, Liu Y, et al. Effects of copper foil as cathode current collector on performance of Li-ion batteries. Corrosion Science and Protection Technology. 2007;**19**(4):265

[6] Jinjin Z, Shuxia R, Yan-liang D, et al. Study on the influence of carbon nanotubes on the mechanical properties of Portland cement. Bulletin of the Chinese Ceramic Society. 2013;**32**(7): 1361-1370

[7] Terrones M. Science and technology of the twenty-first century: Synthesis, properties, and applications of carbon nanotubes. Annual Review of Materials Research. 2003;**33**(1):419-501

[8] Yakobson BI, Avouris P. Mechanical properties of carbon nanotubes. Topics in Applied Physics. 2001:287-327

[9] Anderson RE, Guan J, Ricard M, et al. Multifunctional single-walled carbon nanotube-cellulose composite paper. Journal of Materials Chemistry. 2010; **20**(12):2400-2407

[10] Hu L, Choi JW, Yang Y, et al. Highly conductive paper for energy-storage devices. Proceedings of the National Academy of Sciences. 2009;**106**(51): 21490-21494

[11] Jung R, Kim HS, Kim Y, et al. Electrically conductive transparent papers using multiwalled carbon nanotubes. Journal of Polymer Science Part B: Polymer Physics. 2008;**46**(12): 1235-1242

[12] Tian H, Yang Y, Xie D, et al. A novel flexible capacitive touch pad based on graphene oxide film. Nanoscale. 2013; **5**(3):890-894

[13] Malara F, Manca M, Marco LD, et al. Flexible carbon nanotube-based composite plates as efficient monolithic counter electrodes for dye solar cells. ACS Applied Materials & Interfaces. 2011;**3**(9):3625

[14] Hecht DS, Hu L, Irvin G. Emerging transparent electrodes based on thin films of carbon nanotubes, graphene, and metallic nanostructures. Advanced Materials. 2011;**23**(13):1482-1513

[15] Brennan LJ, Byrne MT, Bari M, et al. Carbon nanomaterials for dye-sensitized solar cell applications: A bright future. Advanced Energy Materials. 2011;**1**(4):472-485

[16] Hu L, Hecht D, Grüner G. Percolation in transparent and conducting carbon nanotube networks. Nano Letters. 2014;**4**(12):2513-2517

[17] Meng Y, Xu XB, Li H, et al. Optimisation of carbon nanotube ink for large-area transparent conducting films fabricated by controllable rod-coating method. Carbon. 2014;**70**(2):103-110

[18] Tenent RC, Barnes TM, Bergeson JD, et al. Ultrasmooth, large-area,

high-uniformity, conductive transparent single-walled-carbon-nanotube films for photovoltaics produced by ultrasonic spraying. Advanced Materials. 2010; **21**(31):3210-3216

[19] Wang J, Liang M, Fang Y, et al. Rod-coating: Towards large-area fabrication of uniform reduced graphene oxide films for flexible touch screens. Advanced Materials. 2012;**24**(21): 2874-2878

[20] Hellstrom SL, Lee HW, Bao Z. Polymer-assisted direct deposition of uniform carbon nanotube bundle networks for high performance transparent electrodes. ACS Nano. 2009;**3**(6):1423-1430

[21] Parekh BB, Fanchini G, Eda G, et al. Improved conductivity of transparent single-wall carbon nanotube thin films via stable postdeposition functionalization. Applied Physics Letters. 2007;**90**(12):2513

[22] Fukaya N, Dong YK, Kishimoto S, et al. One-step sub-10 μm patterning of carbon-nanotube thin films for transparent conductor applications. ACS Nano. 2014;**8**(4):3285-3293

[23] Trung TQ, Ramasundaram S, Hong SW, et al. Flexible and transparent nanocomposite of reduced graphene oxide and P(VDF-TrFE) copolymer for high thermal responsivity in a field-effect transistor. Advanced Functional Materials. 2014; **24**(22):3438-3445

[24] Kim SH, Song W, Jung MW, et al. Carbon nanotube and graphene hybrid thin film for transparent electrodes and field effect transistors. Advanced Materials. 2014;**26**(25):4247-4252

[25] Wu Z, Chen Z, Du X, et al. Transparent, conductive carbon nanotube films. Science. 2004; **305**(5688):1273-1276

[26] Guo P, Song H, Chen X. Electrochemical performance of graphene nanosheets as anode material for lithium-ion batteries. Electrochemistry Communications. 2009;**8**(1):137-142

[27] Cao X, Chuan X, Li S, et al. Hollow silica spheres embedded in a porous carbon matrix and its superior performance as the anode for lithium-ion batteries. Particle & Particle Systems Characterization. 2016;**33**(2): 110-117

[28] Yang Y, Wang Z, Zhou Y, et al. Synthesis of porous Si/graphite/carbon nanotubes@C composites as a practical high-capacity anode for lithium-ion batteries. Materials Letters. 2017;**199**: 84-87

[29] Ma T, Yu X, Li H, et al. High volumetric capacity of hollow structured SnO_2@Si nanospheres for lithium-ion batteries. Nano Letters. 2017;**17**(6):3959

[30] An W, Fu J, Mei S, et al. Dual carbon layers hybridized mesoporous tin hollow spheres for fast-rechargeable and high-stable lithium-ion battery anode. Journal of Materials Chemistry A. 2017;**5**(27)

[31] Jeong MG, Du HL, Islam M, et al. Self-rearrangement of silicon nanoparticles embedded in micro-carbon sphere framework for high-energy and long-life lithium-ion batteries. Nano Letters. 2017;**17**(9)

[32] Iijima S. Helical microtubules of graphitic carbon. Nature. 1991; **354**(6348):56-58

[33] Yoon S, Lee S, Kim S, et al. Carbon nanotube film anodes for flexible lithium ion batteries. Journal of Power Sources. 2015;**279**:495-501

[34] Chen S, Shen L, Peter A, et al. Dual-functionalized double carbon shells

coated silicon nanoparticles for high performance lithium-ion batteries. Advanced Materials. 2017:1605650

[35] Ye H, Xin S, Yin YX, et al. Advanced porous carbon materials for high-efficient lithium metal anodes. Advanced Energy Materials. 2017: 1700530

[36] Luo F, Chu G, Xia X, et al. Thick solid electrolyte interphases grown on silicon nanocone anodes during slow cycling and their negative effects on the performance of Li-ion batteries. Nanoscale. 2015;**7**(17):7651

[37] Wang H, Li Z, Ghosh K, et al. Synthesis of double-walled carbon nanotube films and their field emission properties. Carbon. 2010;**48**(10): 2882-2889

[38] Shinke K, Ando K, Koyama T, et al. Properties of various carbon nanomaterial surfaces in bilirubin adsorption. Colloids & Surfaces B: Biointerfaces. 2010;**77**(1):18-21

[39] Ghosh K, Kumar M, Maruyama T, et al. Micro-structural, electron-spectroscopic and field-emission studies of carbon nitride nanotubes grown from cage-like and linear carbon sources. Carbon. 2009;**47**(6):1565-1575

[40] Beaulieu LY, Hatchard TD, Bonakdarpour A, et al. Reaction of Li with alloy thin films studied by in situ AFM. Journal of the Electrochemical Society. 2003;**150**(11):A1457-A1464

[41] Wada T, Ichitsubo T, Yubuta K, et al. Bulk-nanoporous-silicon negative electrode with extremely high cyclability for lithium-ion batteries prepared using a top-down process. Nano Letters. 2014;**14**:4505-4510

[42] Lv Q, Liu Y, Ma T, et al. Hollow structured silicon anodes with stabilized solid electrolyte interphase film for lithium-ion batteries. ACS Applied Materials & Interfaces. 2015;**7**: 23501-23506

[43] Sun XG, Qiu ZW, Chen L, et al. Industrial synthesis of Whisker carbon nanotubes. Materials Science Forum. Trans Tech Publications Ltd. 2016; **852**:514

[44] Huang X, Yang J, Mao S, et al. Controllable synthesis of hollow Si anode for long-cycle-life lithium-ion batteries. Advanced Materials. 2014; **26**(25):4326-4332

[45] Ma H, Cheng F, Chen J, et al. Nest-like silicon nanospheres for high-capacity lithium storage. Advanced Materials. 2010;**19**(22):4067-4070

[46] Zhou XY, Tang JJ, Yang J, et al. Silicon@carbon hollow core–shell heterostructures novel anode materials for lithium-ion batteries. Electrochimica Acta. 2013;**87**(1):663-668

[47] Sheng L, Jin A, Yu L, et al. A simple and universal method for fabricating linear carbon chains in multiwalled carbon nanotubes. Materials Letters. 2012;**81**(3):222-224

[48] Ghosh K, Kumar M, Maruyama T, et al. Tailoring the field emission property of nitrogen-doped carbon nanotubes by controlling the graphitic/pyridinic substitution. Carbon. 2010; **48**(1):191-200

[49] Ghosh K, Kumar M, Maruyama T, et al. Microstructural, electron-spectroscopic and field-emission studies of carbon nitride nanotubes grown from cage-like and linear carbon sources. Carbon. 2009;**47**:1565-1575

[50] Andoa K, Shinkea K, Yamadaa S, Koyamab T, et al. Fabrication of carbon nanotube sheets and their bilirubin adsorption capacity. Colloids and Surfaces B: Biointerfaces. 2009;**71**(2): 255-259

[51] Kumar M, Ando Y. Camphor—A botanical precursor producing garden of carbon nanotubes. Diamond & Related Materials. 2003;**12**:998-1002

[52] Shimizu T, Abe H, Ando A, et al. Electric transport measurement of a multi-walled carbon nanotube in scanning transmission electron microscope. Physica E: Low-dimensional Systems and Nanostructures. 2004;**24**:37-41

[53] Taniguchi M, Nagao H, Hiramatsu M, et al. Preparation of dense carbon nanotube film using microwave plasma-enhanced chemical vapor deposition. Diamond & Related Materials. 2005; **14**(1):855-858

[54] Endo M et al. Stacking nature of graphene layers in carbon nanotubes and nanofibres. Journal of Physics & Chemistry of Solids. 1997;**58**(11): 1707-1712

Chapter 6

TiO_2 Nanoparticles Prepared by Sol-Gel Method for Anode Application in Lithium-Ion Batteries

Wafaa Nachit, S. Touhtouh, Z. Ramzi, K. Benkhouja and R. Yazami

Abstract

TiO_2 nanoparticles are prepared via sol-gel method using titanium tetraisopropoxide (TTIP) as a precursor under refluxing and controlled pH. It is found that pure anatase phase is obtained with pH 10. Further characterization studies are carried out on pure nanoparticle anatase phase by XRD, SEM, and transmission electron microscope (TEM). Their electrochemical performances as anode material in lithium-ion batteries are investigated by cyclic voltammetry, galvanostatic cycling, and rate capability measurements. A high discharge capacity of 321 mAh/g (vs. 335 mAh/g theoretical) is achieved at 1C rate. After the first galvanostatic charge/discharge cycle, voltage profiles show plateaus at 1.75 and 1.95 V versus Li at discharge and charge, respectively. High Coulombic efficiency (>99%) is maintained after 300 cycles, which makes anatase TiO_2 nanoparticles prepared by sol-gel method, a very promising material for anode application in lithium rechargeable batteries.

Keywords: TiO_2 nanoparticles, anode, lithium-ion batteries, electrochemical characteristics

1. Introduction

Nanostructured titania (NST) TiO_2 is a very promising alternative anode material compared with the commercialized lithium titanate $Li_4Ti_{15}O_{12}$ and graphite materials for lithium-ion battery application in particular when fast charging is required [1–3]. Fast charging while keeping long battery operational life is one of the most challenging tasks in lithium-ion batteries nowadays. Both mobile electronics and electro-mobility sectors need fast charging to compensate for relatively low energy density which impacts the operation time and driving range. One serious option to cope with charging time limitations resides in developing new materials able to store large amounts of lithium in a very short time. Nanomaterials, including NST, are among the most promising in this regard.

Good performances of NST result from their outstanding characteristics such as:

i. A low volumetric change upon lithium-ion intercalation and de-intercalation providing a high crystal structure stability and long cycle life

ii. A fast diffusivity of lithium ions within the TiO_2 structure providing fast charging and high power discharge

iii. A practical low operating voltage

iv. An enhanced safety

Several methods have been used for the synthesis of NST, including hydrothermal [4–10], sol-gel [5–11], hydrolysis [12, 13], atomic layer deposition [14], and anode deposition [15], for the most reported in the literature. According to the synthesis method and to the post-thermal treatment conditions, NST of different morphologies are obtained such as nanorods [4, 17], nanotubes [5–6, 8–9, 14, 17], nanofibers/nanowires [11–20], nanoparticles [6, 7, 10], nanocomposites [12, 21], nanofilms, and nanosheets [15, 22].

Several polymorphs of TiO_2 are known such as tetragonal rutile and anatase, orthorhombic brookite, and monoclinic TiO_2(B) [23]. Anatase and TiO_2(B) have open-channel structure and show high cycle capacity compared to rutile TiO_2 [24]. Anatase has edge-sharing TiO_6 octahedra so that planar double chains are formed with a tetragonal body-centered space group ($I4_1$/amd) [25].

Li-ion diffusion in anatase occurs along the octahedral interstitial sites up to $Li_{0.5}TiO_2$ composition. Then the tetragonal structure reversibly transforms to the orthorhombic phase [26–30]. Anatase TiO_2 anode material reversibly uptakes 0.5 Li per TiO_2 formula with lithium insertion/extraction taking place at relatively low voltage and showing fast kinetics [29, 30]. However, 0.5 Li per TiO_2 corresponds to half of theoretical capacity of 335 mAh/g [31–33]. Enhancing practical capacity in NST should make this material more attractive for anode application in lithium-ion batteries [34].

When wet chemistry route is used to prepare NST, the medium pH is of a significant importance in the crystal structure and morphology of the final product [32]. Syntheses of TiO_2 anatase were mostly performed in acidic media usually at pH greater than 4.5 [33]. To the authors' knowledge, there has been no report on sol-gel preparation of NST in alkaline pH.

In this chapter, we show that unlike in acidic pH, which lead to mixed phases, pure anatase NST is obtained via sol-gel at pH 10. NST prepared at different pH are analyzed by XRD, transmission electron microscope (TEM), and SEM. The electrochemical properties of pure anatase phase NST are investigated by cyclic voltammetry and galvanostatic cycling, including rate capability tests in view of its application as anode material in lithium-ion batteries.

2. Experimental

2.1 Sample preparation

NST are prepared via sol-gel method using titanium tetraisopropoxide (TTIP), distilled water, and 2-propanol. Hydrochloric acid and sodium hydroxide are mixed in 24 mL of 2-propanol to fix the pH to 1, 4, 10, and 13 and to control the hydrolysis process of the solution. 24 mL of TTIP is then slowly added to the mixture, followed by refluxing at 70°C to obtain the sol. 10 mL of distilled water is added to the solution and left for 1 h under stir heating.

The obtained gel is dried at 70°C in air and then calcined at 500°C in air to produce NST powders.

2.2 Characterization

NST are characterized by X-ray diffractometry (XRD) using a PR-XPERT diffractometer (PW3064) with CuK_α radiation (λ = 1.5406 Å), operating at 45 kV and 40 mA in 2θ angular range of 10–70°. The morphology and particle size were examined by transmission electron microscope (TEM, JEOL, JEM-2010) and by field-emission scanning electron microscopy (FESEM, JEOL JSM-7600F, 5 kV).

2.3 Electrochemical measurements

The anode is prepared by mixing NST prepared at pH 10 as active material, acetylene carbon black as electronic conductor, and polyvinylidene fluoride (PVDF) as a binder in the weight ratio of 80:10:10 with N-methyl-2-pyrrolidone (NMP) to form a slurry. The slurry was coated on a copper foil by doctor blade and then vacuum dried at 110°C for 6 h. The anode loading is ~0.8 mg cm^{-2}. Lithium half-cells are assembled in a glove box under ultra-dry argon atmosphere (<1 ppm H_2O and O_2). Lithium metal was used as the counter and reference electrode. The electrolyte consisted of 1.0 M $LiPF_6$ solution in ethylene carbonate (EC) and diethyl carbonate (DEC) at a volumetric ratio of 1:1. The separator is microporous polyethylene. Li/NST half-cells are investigated by cyclic voltammetry at 0.2 mV/s sweeping rate and by galvanostatic cycling at 1C–20 C rates, all between 1.0 V and 3.0 V versus Li.

3. Results and discussion

3.1 Physical characterizations

Figure 1 shows the XRD patterns of TiO_2 nanoparticles prepared at pH 1, 4, 10, and 13 and calcined at 500°C. X-ray patterns of anatase, rutile, and brookite are shown for reference.

The sample prepared at pH 1 consists mostly of the rutile phase. All other samples are anatase with brookite impurities at pH 4 and 13. TiO_2 nanoparticles synthesized at pH 10 consist of highly crystalline anatase phase (00-004-0477). **Table 1** summarizes the XRD data of TiO_2 nanoparticles synthesized at pH 10 including diffraction peak 2θ, interlayer spacing "d," and hkl indexes. **Table 2** gives the crystallographic characteristics of TiO_2 anatase phase prepared at pH 10 as derived from the XRD data Rietveld refinement. Average crystallite sizes of nanoparticles synthesized at different pH are reported in **Table 3** using Scherrer's equation. The particle size is in ~20–30 nm range with no obvious trends versus pH.

Figure 2 shows TEM images of TiO_2 nanoparticles prepared at pH 10 at different magnifications. **Figure 2a** and **b** shows primary particles with relatively narrow size distribution. High-resolution TEM of **Figure 2c** shows the polycrystalline nature of the TiO_2 nanoparticles with well-resolved layers such as those assigned to the orientation corresponding to an interlayer of about 0.36 nm as indicated by an arrow.

TEM image analysis allows particle-size distribution to be determined as shown in **Figure 3**. The particle-size distribution is built on a discrete set of clearly identifiable grains, and, therefore, it is not necessarily representative of the overall description of the sample. This is especially true for grain sizes close to the distribution tails where the probability of finding a very large or a very small grain under the TEM objective is rather low. The average size of about 17 nm is slightly lower than the one estimated from the XRD data.

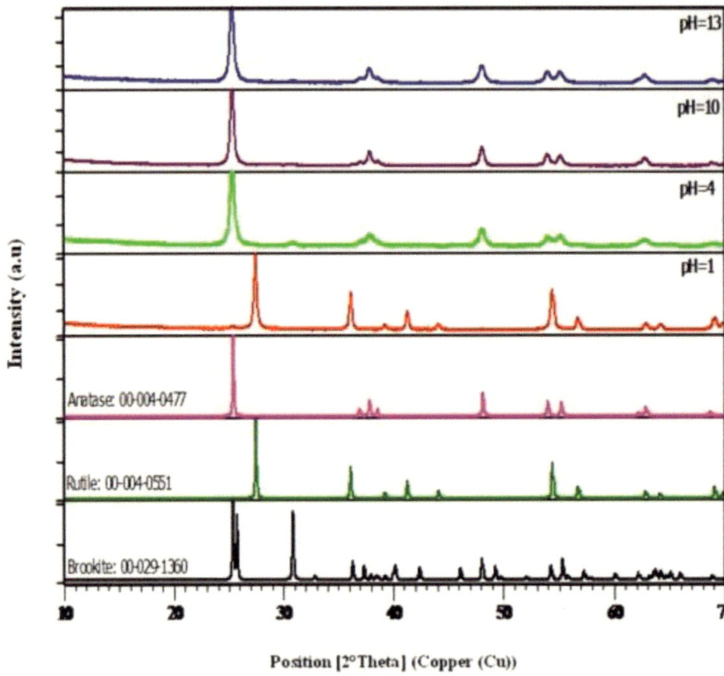

Figure 1.
XRD patterns of titania nanoparticles made at different pH. Anatase, rutile, and brookite spectra are also shown for reference.

2θ (°)	d (nm)	(hkl)
25.31	0.3518	(101)
36.98	0.2430	(103)
37.82	0.2378	(004)
38.61	0.2331	(112)
48.05	0.1893	(200)
53.92	0.1700	(105)
55.09	0.1637	(221)
62.13	0.1494	(213)
62.70	0.1481	(204)
62.80	0.1479	(116)

Table 1.
Indexation of XRD peaks on TiO_2 nanoparticles obtained at pH 10.

SEM images of NST pH 10 are displayed in **Figure 4a** and **b**. High magnification **Figure 4a** shows primary particles with narrow size distribution aggregate into secondary particles of about 150–600 nm in size, whereas lower magnification **Figure 4b** shows agglomerated particles remain basically submicron in size.

3.2 Electrochemical characterizations

Figure 1a shows cyclic voltammograms of the Li/NST of pH 10 half-cell at 0.2 mV/s sweeping rate for 5 cycles. In the first cycle, two reduction peaks appear at 2.15 V and at 1.75 V, respectively, whereas only one re-oxidation peak is present at 2.10 V.

Phase	α β γ	Lattice parameters	
Anatase (Tetragonal I41)	α = β = γ = 90°	a = b = 3.783	c = 9.5190

Table 2.
Characteristics of anatase crystal structure.

pH	Phase	D (nm)
1	Anatase-rutile	32.0
4	Anatase-brookite	21.3
10	Anatase	26.2
13	Anatase-brookite	21.3

Table 3.
Crystallite size of TiO_2 nanoparticles synthesized at different pH.

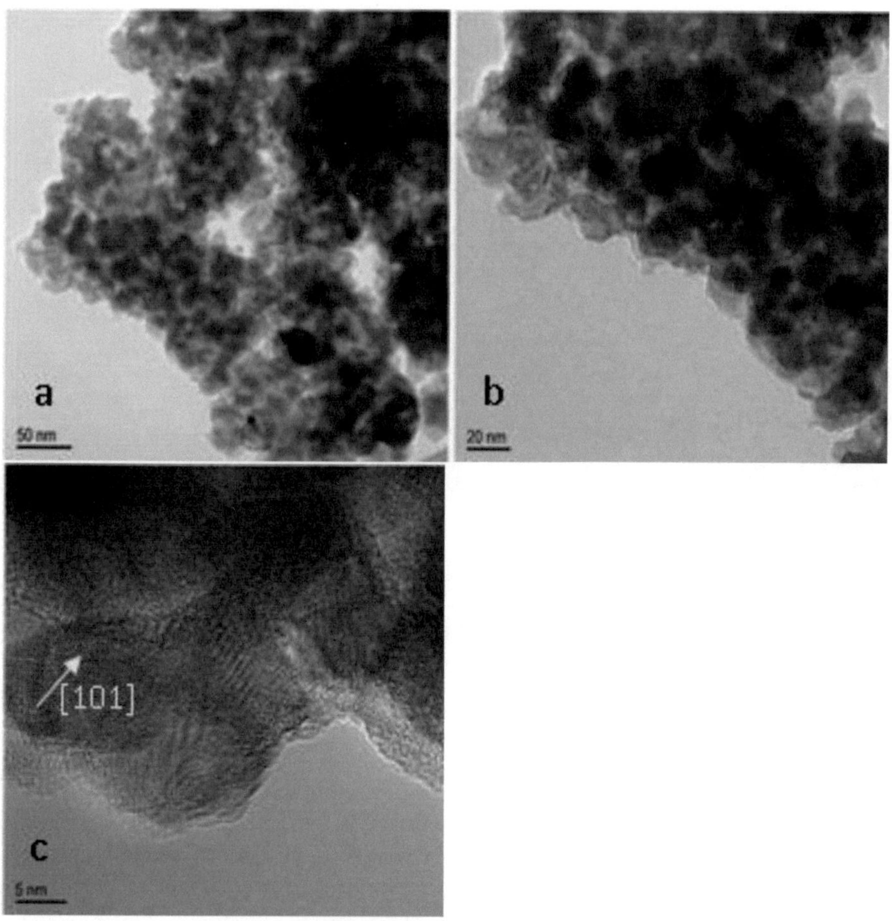

Figure 2.
TEM images at different magnifications of TiO_2 nanoparticles prepared at pH 10. (a, b) Image analysis enables size distribution to be achieved. HR-TEM image of (c) shows polycrystalline structure with some preferred layer orientation such as with interlayer spacing of about 3.6 nm.

In the following 4 cycles, only the reduction peak at 1.75 V and re-oxidation peak at 2.10 V appeared. The first reduction peak at 2.15 V accounts for about 50% of the total discharge capacity of about 320 mAh/g, which is close to theoretical capacity

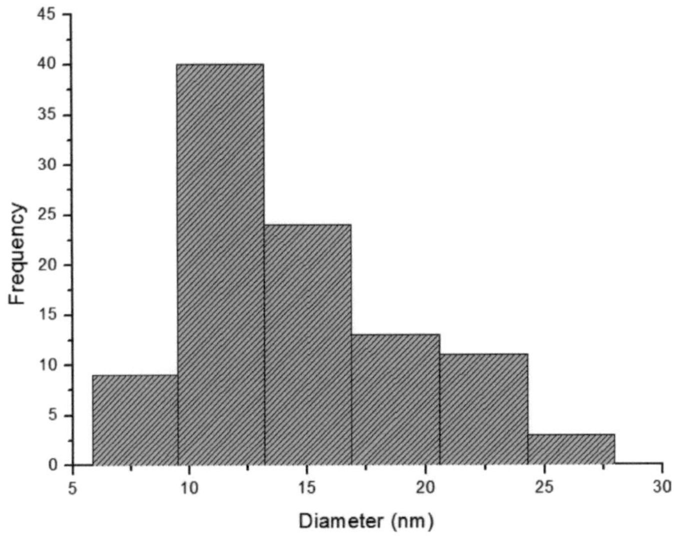

Figure 3.
Grain size distribution for the TiO_2 at pH 10 from the TEM image analysis.

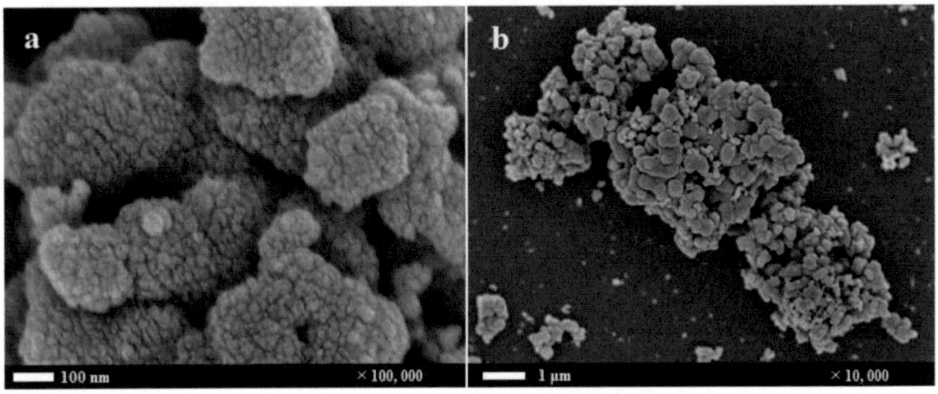

Figure 4.
SEM images of TiO_2 nanoparticles synthesized at pH 10.

of 335 mAh/g of TiO_2 corresponding to 1 Li/TiO_2 ratio. The second reduction peak at 1.75 V accounts for the second half of the total capacity and is matched by the re-oxidation peak at 2.1 V. This situation remained almost unchanged during the four following cycles, that is, the capacity of the reduction peak at 1.75 V matches the one of the re-oxidation peaks at 2.1 V of about 140 mAh/g.

The data shown above are confirmed during the two first galvanostatic cycles of the Li/NST cell displayed in **Figure 5b**. The first discharge profile shows two semi-plateaus at about 2.15 and 1.75 V, respectively, whereas the charge profile shows a main plateau at about 2.0 V. The first discharge semi-plateau accounts for nearly half of the total discharge capacity. The second half discharge capacity is matched by the charge capacity.

Cyclic voltammetry and galvanostatic data suggest a two-step electrode reaction process:

1. Irreversible formation of $Li_{0.5}TiO_2$ at ~ 2.15 V average voltage.

2. Reversible formation of Li_1TiO_2 from $Li_{0.5}TiO_2$ at ~ 1.75 V. During re-oxidation $Li_{0.5}TiO_2$ is formed from Li_1TiO_2 at about 2.1 V.

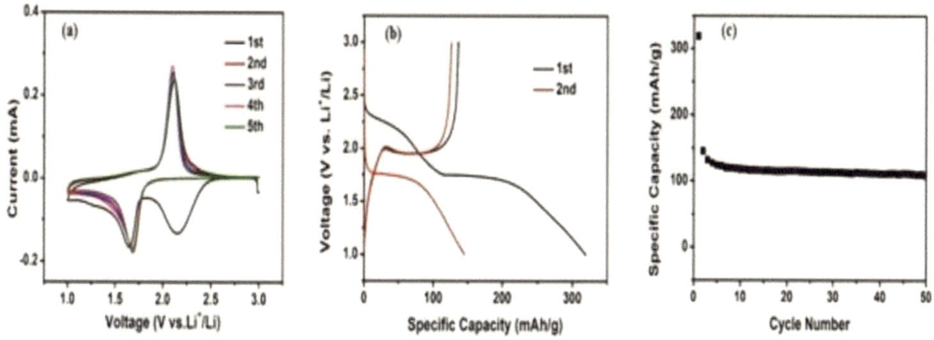

Figure 5.
Electrochemical tests performed on Li/TiO$_2$ (pH 10) half-cells. (a) Cyclic voltammograms (CVs) at a scan rate of 0.2 mV/s for the first 5 cycles within a potential range from 1 to 3.0 V (vs. Li/Li$^+$), (b) discharge-charge capacity of the first and second cycles at a specific current of 175 mA/g (1C), and (c) cycle capacity profile at 1C rate for 50 cycles.

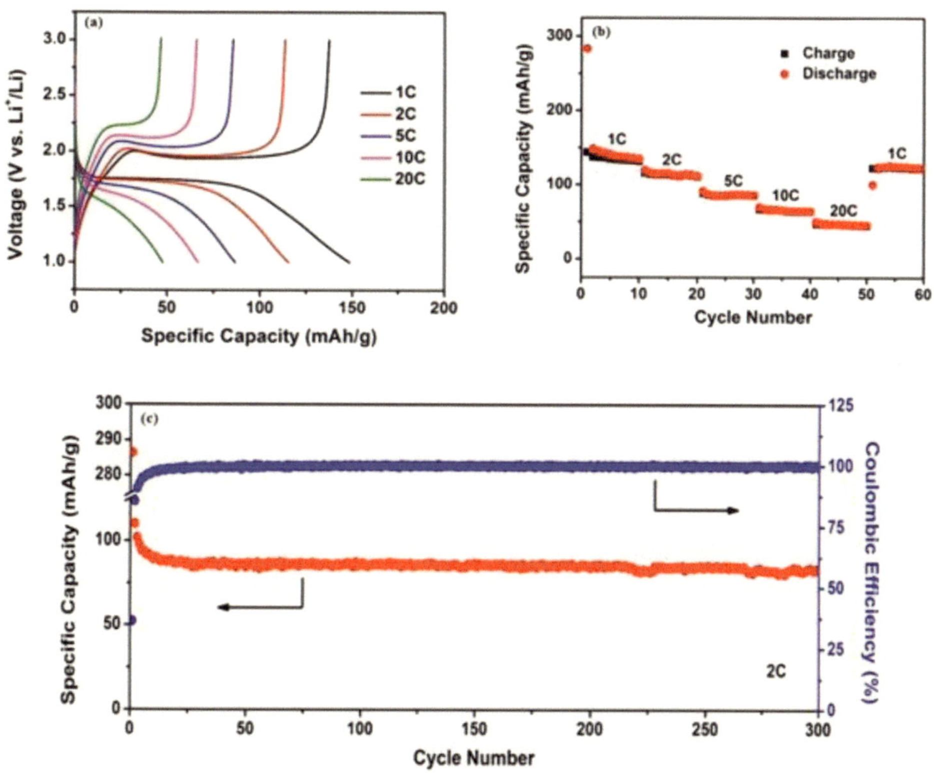

Figure 6.
Rate capability (1C–20C) tests performed on Li/TiO$_2$ (pH 10) half-cells. (a) Discharge-charge voltage profiles, (b) discharge-charge capacity profile, and (c) cycle capacity and Coulombic efficiency profiles at 2C rate.

The formation of lithium-rich Li_1TiO_2 is supported by both theoretical and experimental data [30–35]. In the anatase phase, orthorhombic $Li_{0.5}TiO_2$ and distorted rock-salt structure Li_1TiO_2 in which lithium cation occupy the octahedral sites have been stabilized due to kinetic effects. Metastability has been reported in Li-poor TiO_2 [30]. Accordingly, our data are consistent with reversible lithium intercalation in the stoichiometry range $0.5 < x < 1$ in anatase Li_xTiO_2, which corresponds to a theoretical capacity of 167.5 mAh/g. The cycle capacity profile in

Figure 5c shows a plateau at about 120 mAh/g at 350 mA/g rate. A lower capacity may be due to limitation in lithium uptake to $Li_{0.85}TiO_2$ and $Li_{0.7}TiO_2$ as reported in [36, 37].

The intercalation mechanism of Li ions into anatase has been studied in a variety of experimental and theoretical works [38]. The lithium-ion intercalation can be described by the following Eq. (1):

$$TiO_2 + xLi^+ + xe^- \rightarrow Li_xTiO_2 \quad (1)$$

TiO_2 maximum theoretical capacity is 335 mAh/g [39] which corresponds to $x = 1$ and to the complete reduction of Ti^{4+} to Ti^{3+}. This makes TiO_2 a highly competitive alternative to graphite anodes [40].

Practical capacity of anatase TiO_2 is about 200 mAh/g associated with 0.6 Li mole reversible intercalation at 1.78 V versus Li^+/Li [41]. It has been demonstrated that Li_1TiO_2 composition can be obtained from anatase TiO_2 at high temperature [30] or with particle size below 7 nm [42].

The discharge-charge profiles at current rates ranging from 1.0C to 20C and corresponding rate capability profiles are depicted in **Figure 6a** and **b**, respectively. It is worth noticing that after 10 cycles at 20C rate where capacity decreased to about 50 mAh/g, the capacity at 1C rate was recovered during the following 10 cycles to about 137 mAh/g. This indicates a high stability of the anatase TiO_2 nanoparticles based anode toward high rate cycling.

Figure 6c shows the cycle capacity profile at 2C rate for 300 cycles together with Coulombic efficiency. The latter stabilizes close to unity after the first 5 cycles, whereas the capacity gradually dropped to 82.4 mAh/g after 300 cycles indicating the high capacity retention and good cycling stability.

4. Conclusion

TiO_2 nanoparticles are synthesized by simple sol-gel method at different pH values and found to have particle size of 10–30 nm. XRD measurements show that the sample prepared at pH 10 crystallized to pure anatase phase. The TEM and the SEM images confirm the formation of nanoparticles with diameters of 10–30 nm. The TiO_2 nanoparticles prepared at pH 10 exhibit initial capacity of 321 mAh/g and a remarkable high Coulombic efficiency close to unity after 300 cycles at 2C rate. However, the first cycle efficiency is relatively low perhaps due to inherent metastability of Li-poor phases at the early stage of lithium intercalation. More work has currently been performed to improve the first cycle efficiency, therefore making anatase nanoparticles based anode more practical in rechargeable lithium batteries with fast charge and high lithium storage capacity. Among possible routes considered to improve the NTO initial cycle efficiency is the incorporation of a few amounts of lithium during sol-gel process so as to achieve a pre-lithiated phase making it easier to add more lithium during the first cathodic process.

Author details

Wafaa Nachit[1,2], S. Touhtouh[2], Z. Ramzi[1,2], K. Benkhouja[1] and R. Yazami[3]*

1 Laboratory of Coordination and Analytical Chemistry (LCCA (E2M)), Faculty of Sciences, Department of Chemistry, Chouaib Doukkali University, El Jadida, Morocco

2 Engineering Sciences Laboratory for Energy (LabSIPE), National School of Applied Sciences (ENSA), Chouaib Doukkali University, El Jadida, Morocco

3 Energy Research Institute, Nanyang Technological University, Singapore, Singapore

*Address all correspondence to: rachid@ntu.edu.sg and nachit.wafaa@gmail.com

References

[1] Wang C, Wu LX, Wang H, Zuo WH, Li YY, Liu JP. Fabrication and shell optimization of synergistic TiO_2-MoO_3 core-shell nanowire array anode for high energy and power density lithium-ion batteries. Advanced Functional Materials. 2015;**15**(23):3524-3533

[2] Wei W, Oltean G, Tai CW, Edstrom K, Bjorefors F, Nyholm L. High energy and power density TiO_2 nanotube electrodes for 3D Li-ion microbatteries. Journal of Materials Chemistry. 2013;**1**(28):8160-8169

[3] Ortiz GF, Hanzu I, Lavela P, Knauth P, Tirado JL, Djenizian T. Nanoarchitectured TiO_2/SnO: A future negative electrode for high power density Li-Ion microbatteries. Chemistry of Materials. 2010;**22**(5):1926-1932

[4] Liu YB, Ding TL, Shen DL, Dou J, Wei MD. Hierarchically structural TiO_2 nanorods composed of rutile core and anatase shell as a durable anode material for lithium-ion intercalation. Journal of Electroanalytical Chemistry. 2017;**804**:87-91

[5] Fang J, Liu WW, Yu F, Qin FR, Wang MR, Zhang K, et al. Fe, S co-doped anatase TiO_2 nanotubes as anodes with improved electrochemical performance for lithium ion batteries. RSC Advances. 2016;**6**(74):70133-70,140

[6] Tang YK, Liu L, Zhao HY, Jia DZ, Xie XL, Zhang Y, et al. Anatase/rutile titania anchored carbon nanotube porous nanocomposites as superior anodes for lithium ion batteries. CrystEngComm. 2016;**18**(24):4489-4494

[7] Sanad MMS, Rashad MM, Powers K. Enhancement of the electrochemical performance of hydrothermally prepared anatase nanoparticles for optimal use as high capacity anode materials in lithium ion batteries (LIBs). Applied Physics A: Materials Science and Processing. 2015;**118**(2):665-674

[8] Choi MG, Lee YG, Song SW, Kim KM. Anode properties of titanium oxide nanotube and graphite composites for lithium-ion batteries. Electrochimica Acta. 2010;**55**(20):5975-5983

[9] Xu JW, Jia CH, Cao B, Zhang WF. Electrochemical properties of anatase TiO_2 nanotubes as an anode material for lithium-ion batteries. Electrochimica Acta. 2007;**52**(28):8044-8047

[10] Oh SW, Park SH, Sun YK. Hydrothermal synthesis of nano-sized anatase TiO_2 powders for lithium secondary anode materials. Journal of Power Sources. 2006;**161**(2):1314-1318

[11] Lee S, Eom W, Park H, Han TH. High-temperature stable anatase titanium oxide nanofibers for lithium-ion battery anodes. ACS Applied Materials and Interfaces. 2017;**9**(30):25332-25,338

[12] Liu JL, Feng HB, Jiang JB, Qian D, Li JH, Peng SJ, et al. Anatase-TiO_2/CNTs nanocomposite as a superior high-rate anode material for lithium-ion batteries. Journal of Alloys and Compounds. 2014;**603**:144-148

[13] Xu W, Wang ZY, Guo ZD, Liu Y, Zhou NZ, Niu B, et al. Nanoporous anatase TiO_2/single-wall carbon nanohorns composite as superior anode for lithium ion batteries. Journal of Power Sources. 2013;**232**:193-198

[14] Panda SK, Yoon Y, Jung HS, Yoon WS, Shin H. Nanoscale size effect of titania (anatase) nanotubes with uniform wall thickness as high performance anode for lithium-ion secondary battery. Journal of Power Sources. 2012;**204**:162-167

[15] Wu MS, Wang MJ, Jow JJ, Yang WD, Hsieh CY, Tsai HM. Electrochemical fabrication of anatase TiO(2) nanostructure as an anode material for aqueous lithium-ion batteries. Journal of Power Sources. 2008;**185**(2):1420-1424

[16] Kim J, Cho J. Rate characteristics of anatase TiO_2 nanotubes and nanorods for lithium battery anode materials at room temperature. Journal of the Electrochemical Society. 2017;**154**(6):A542-A546

[17] Zhang MH, Yin KB, Hood ZD, Bi ZH, Bridges CA, Dai S, et al. In situ TEM observation of the electrochemical lithiation of N-doped anatase TiO_2 nanotubes as anodes for lithium-ion batteries. Journal of Materials Chemistry A. 2017;**5**(39):20651-20,657

[18] Luo Y, Li J, Huang JG. Bioinspired hierarchical nanofibrous silver-nanoparticle/anatase-rutile-titania composite as an anode material for lithium-ion batteries. Langmuir. 2016;**32**(47):12338-12,343

[19] Yang XJ, Teng DH, Liu BX, Yu YH, Yang XP. Nanosized anatase titanium dioxide loaded porous carbon nanofiber webs as anode materials for lithium-ion batteries. Electrochemistry Communications. 2011;**139**(10):1098-1101

[20] Wu FX, Li XH, Wang ZX, Guo HJ, Wu L, Xiong XH, et al. A novel method to synthesize anatase TiO_2 nanowires as an anode material for lithium-ion batteries. Journal of Alloys and Compounds. 2011;**509**(18):3711-3715

[21] Lai C, Li CGRY, Dou Y, Gao XP. Mesoporous polyaniline or polypyrrole/anatase TiO_2 nanocomposite as anode materials for lithium-ion batteries. Electrochimica Acta. 2010;**55**(15):4567-4572

[22] Cheng XL, Hu M, Huang R, Jiang JS. HF-free synthesis of anatase TiO_2 nanosheets with largely exposed and clean {001} facets and their enhanced rate performance as anodes of lithium-ion battery. ACS Applied Materials and Interfaces. 2014;**6**(2):19176-19,183

[23] Roy P, Srivastava SK. Nanostructured anode materials for lithium ion batteries. Journal of Materials Chemistry A. 2015;**3**(6):2454-2484

[24] Wang XD, Li ZD, Shi J, Yu YH. One-dimensional titanium dioxide nanomaterials: Nanowires, nanorods, and nanobelts. Chemical Reviews. 2014;**114**(19):9346-9384

[25] Ortiz GF, Hanzu I, Djenizian T, Lavela P, Tirado JL, Knauth P. Alternative Li-ion battery electrode based on self-organized titania nanotubes. Chemistry of Materials. 2009;**21**(1):63-67

[26] Sudant G, Baudrin E, Larcher D, Tarascon JM. Electrochemical lithium reactivity with nanotextured anatase-type TiO_2. Journal of Materials Chemistry. 2005;**15**(12):1263-1269

[27] Wagemaker M, Borghols WJH, van Eck ERH, Kentgens APM, Kearley GL, Mulder FM. The influence of size on phase morphology and Li-ion mobility in nanosized lithiated anatase TiO_2. Chemistry--A European Journal. 2007;**13**(7):2023-2028

[28] Wagemaker M, Van Well AA, Kearley GJ, Mulder FM. The life and times of lithium in anatase TiO_2. Solid State Ionics. 2004;**175**(1):191-193

[29] He YB, Liu M, Xu ZL, Zhang B, Li B, Kang F, et al. Li-ion reaction to improve the rate performance of nanoporous anatase TiO_2 anodes. Energy Technology. 2013;**1**(11):668-674

[30] Macklin WJ, Neat RJ. Performance of titanium dioxide-based cathodes in

a lithium polymer electrolyte cell. Solid State Ionics. 1992;**53**:694-700

[31] Zhu GN, Wang YG, Xia YY. Ti-based compounds as anode materials for Li-ion batteries. Energy and Environmental Science. 2012;**5**(5):6652-6667

[32] Zhou ZY, Tian N, Li JT, Broadwell I, Sun SG. Nanomaterials of high surface energy with exceptional properties in catalysis and energy storage. Chemical Society Reviews. 2011;**40**(7):4167-4185

[33] Bavykin DV, Friedrich JM, Walsh FC. Protonated titanates and TiO_2 nanostructured materials: Synthesis, properties, and applications. Advanced Materials. 2006;**18**(2):2807-2824

[34] Zakharova GS, Jahne C, Popa A, Taschner C, Gemming T, Leonhardt A, et al. Anatase nanotubes as an electrode material for lithium-ion batteries. Physical Chemistry C. 2012;**116**(15):8714-8720

[35] Wagemaker M, Kentgens APM, Mulder FM. Equilibrium lithium transport between nanocrystalline phases in intercalated TiO_2 anatase. Nature. 2002;**418**(6896):397-399

[36] Hu YS, Kienle L, Guo YG, Maier J. High lithium electroactivity of nanometer-sized rutile TiO_2. Advanced Materials. 2006;**18**(11):1421-1426

[37] Murphy DW, Greenblatt M, Zahurak SM, Cava RJ, Waszczak JV, Hull GW, et al. Lithium insertion in anatase - a new route to the spinel $LiTi_2O_4$. Chemischer Informationsdienst. 1982;**19**(4-5):441-449

[38] Wagemaker M, Kearley GJ, van Well AA, Mulka H, Mulder FM. Multiple li positions inside oxygen octahedra in lithiated TiO_2 anatase. Journal of the American Chemical Society. 2003;**125**(3):840-848

[39] Jiang C, Zhang J. Nanoengineering titania for high rate lithium storage: A review. Journal of Materials Science and Technology. 2013;**29**(2):97-122

[40] Fröschl T, Hörmann U, Kubiak P, Kucerova G, Pfanzelt M, Weiss CK, et al. Nanoengineering titania for high rate lithium storage: A review. Journal of Royal Society of Chemistry. 2012;**41**(15):5313-5360

[41] Kavan L, Kalbac M, Zukalova M, Exnar I, Lorenzen V, Nesper R, et al. Lithium storage in nanostructured TiO_2 made by hydrothermal growth. Chemistry of Materials. 2004;**16**(3):477-485

[42] Lafont U, Carta D, Mountjoy G, Chadwick AV, Kelder EM. In situ structural changes upon electrochemical lithium insertion in nanosized anatalse TiO_2. Physical Chemistry C. 2010;**114**(2):1372-1378

Section 5

Thin Film for Solid-Electrolyte

Chapter 7

Surface Chemical Analysis of Solid-Electrolyte Interphase Layer on Germanium Thin Films and the Effect of Vinylene Carbonate Electrolyte Additive

Silpasree S. Jayasree, Shantikumar Nair and Dhamodaran Santhanagopalan

Abstract

Germanium thin-film anodes for Li-ion battery applications are the focus of the present work. As part of this chapter, we shall briefly review the use of germanium thin films in Li-ion batteries, and subsequently, new results pertaining to the effect of vinylene carbonate (VC) as electrolyte additive on the electrochemical performance are presented. We have used cyclic voltammetry, galvanostatic charge-discharge and electrochemical impedance spectroscopy to investigate the performance. Thin-film electrode performance with 0 wt. %, 5 wt. %, and 10 wt.% VC as electrolyte additive was compared to understand the role of additive's concentration. The cell with 5 wt.% VC as electrolyte additive exhibited best performance with high specific capacity of 975 mAh/g, with a retention of 94 and 99% Coulombic efficiency at the end of 100 cycles. Ex situ surface chemical analysis of the solid-electrolyte interphase (SEI) layer has been studied in detail using X-ray photoelectron spectroscopy and correlated with the electrochemical performance.

Keywords: surface chemistry, SEI layer, germanium thin films, electrolyte additives, vinylene carbonate

1. Introduction

With the ever-increasing demand for energy and storage, the need for lithium-ion batteries (LIB) with more safety, high-energy density, power density, and miniaturized size has become critical. Studies have focused on the development of new potential anode materials that could substitute for the commercialized graphite [1, 2]. The group (IV) elements (silicon, germanium, tin) undergoing the mechanism of alloying with lithium have gained attention owing to their high theoretical capacities (4200, 1600, and 994 mAh/g, respectively). Alloying materials have low working potential compared to that of conversion and most of the insertion metal oxides, which is the key parameter for high-energy density full-cell applications [3–5]. Among group (IV) elements, the Li-Si system gained attention for its large gravimetric capacity. High-power application of Si anode is bottlenecked by the huge volume expansion (>300%) associated with alloying/dealloying and by the slow lithium-ion transport. Pulverization of electrodes and crack formation on the active layer leads to rapid capacity fading and poor rate capability over cycling [6–8].

Germanium has gained attention despite its high cost. It offers a high theoretical capacity of 1384 mAh/g ($Li_{15}Ge_4$) with improved kinetics, and it also possesses a volumetric capacity almost equal to Si (7366 Ah/L for $Li_{15}Ge_4$ and 8334 Ah/L for $Li_{15}Si_4$). Lithium-ion diffusivity is higher in Ge, i.e., ~400 times the diffusivity in Si at room temperature, and germanium exhibits 10^4 times higher electrical conductivity than Si [9–11].

Despite of all these attractive advantages, Ge has a few limitations that hinder its commercialization. Some unwanted side reactions between active material and electrolyte result in solid-electrolyte interphase (SEI) layer formation that comprises of lithium carbonates, lithium alkoxides, etc. The pulverization of material occurring due to the drastic volume change during charge/discharge generates an unstable SEI layer on electrode surface, and it also has been reported that diffusion length gets increased because of the aggregation of active particles. This results in increased cell impedance and attenuation of capacity and Coulombic efficiency [12, 13].

Many strategies have been adopted so far to minimize the volume expansion and to improve the electrochemical performance. Nanoscale Ge is considered as an effective method compared to bulk Ge because of the low volume expansion and less aggregation of materials. Increased surface area in nanoscale enhances the ion transport at the electrode/electrolyte interface and also decreases the diffusion length. Several groups investigated and reported the increased performance of Ge nanoparticles [13], nanocrystals [3], nanowires [2], and nanotubes [14] compared to bulk. The importance of surface coating (<1 μm) and alloy formation has been reported by some groups [11]. It shows a reduction in volume expansion and increased electrochemical performance. Fabrication and implementation of Ge thin films (<250 nm thickness) can be treated as a promising solution for attaining stability, rate capability, and long life cycle because of its low volume expansion. Studies reported the improved performance of Ge films in terms of rate capability, electronic and ionic conductivity, retention, etc. [15, 16]. Silicon thin-film electrodes suffer crack formation and loss of electrical contact with the current collector. In contrast, even though crack formation occurs on the surface of the germanium electrode, no lack of electrical contact between material and current collector was observed and hence resulted in superior rate capability of Ge thin films [17]. Electrolyte additives accentuate the electrochemical performance. Vinylene carbonate (VC), fluoroethylene carbonate (FEC), vinyl ethylene carbonate (VEC), etc. are the commonly used additives. These additives will reduce and form stable layer on the electrode surface at a high potential. The surface layer thus formed will allow Li-ion transport through it and inhibit further electrolyte decomposition. SEI layer components depend on the functional groups present in the additive [18, 19].

Herein, we report the fabrication of Ge thin films by thermal evaporation process and study of its structural and electrochemical properties with/without the electrolyte additive, VC. Surface chemical analysis of SEI layer formation on Ge thin films was evaluated in detail to correlate with the electrochemical cycling stability.

2. Brief review on germanium thin-film-based Li-ion battery electrodes

Many groups had studied and reported the potential of germanium thin films as anode for LIB applications. Films were developed via thermal evaporation [6, 12, 17], electron-beam evaporation [8, 15, 16, 20], sputtering [21–24], electrodeposition [25], and alloying [26–28].

Graetz et al. [6] did a comparative study between Ge nanocrystals (12 nm mean diameter) and amorphous thin films (60–250 nm thickness) as anodes for LIB. Nanocrystals were obtained by ballistic deposition, and amorphous film fabrication was done via thermal evaporation. In Ge nanocrystals, different Li-Ge crystalline phases were formed upon lithiation that ended up with severe capacity loss after a few cycles. It is because of the material decrepitating caused by increased stress during cycling. In case of a-Ge thin film, partition of film occurs due to the crack formation and propagation. These island-like partitions accommodate the volume change during cycling, ensuring proper electric contact between substrate and material. Upon lithiation, a-Ge became crystalline, and it remains as such during further delithiation. It delivered a capacity of 1.7 Ah/g over 62 cycles without any capacity loss, whereas Ge nanocrystals exhibited a capacity of 1.4 Ah/g with 60% retention at the end of 50 cycles. Thin-film electrodes showed rate capability up to 1000 C with very less reduction in capacity. They suggest that the absence of defects, increased surface area, and reduction in diffusion length resulted in enhanced performance of a-Ge thin films.

Germanium thin films were fabricated and coupled with single wall carbon nanotube (SWCNT) that acts as current collector. Dileo et al. [8] introduced a method to increase energy density of battery by the removal of inactive current collectors. They deposited Ge on SWCNT via electron-beam evaporation with different loadings of 10, 20, and 40%. Ge-SWCNT with 40 wt.% Ge exhibited an increased capacity of 800 mAh/g. Full cells were developed with 40 wt.% Ge-SWCNT as anode and commercialized $LiCoO_2$ as cathode, also used mesocarbon microbead (MCMB) as anode for comparison. Ge-SWCNT/$LiCoO_2$ full cell maintained a constant discharge voltage of 3.35 V. Ge-SWCNT showed an improved anode energy density of 1600 Wh/kg compared to MCMB (670 Wh/kg).

Bagetto et al. [17] presented the electrochemical and in situ X-ray diffraction study of Ge thin films for the better understanding of Li-ion insertion/deinsertion mechanism. Upon complete lithiation, both the electrodes got transformed to crystalline $Li_{15}Ge_4$ leading to maximum storage capacity for the specified composition. The rate capability test on evaporated films was conducted during delithiation. It was discharged from 0.1C to a high rate of 100C. Eve at 100C, it delivered high capacity that indicates the proper Li diffusion into the evaporated Ge films. Insertion reaction was similar in both the films, whereas during deinsertion they observed a slight difference in the peak position. They were not able to point a proper reason for the difference in reaction of Li with both the films, hence recommended the need of further investigations. Impedance analysis and electrochemical equilibrium revealed the potential of evaporated films as a promising anode material than sputtered one.

Investigation on Ge thin films developed via radio frequency magnetron sputtering was done by Laforge et al. [21]. Doped and non-doped Ge targets were used for the deposition. Effect of doping concentration and electrode thickness was reported in detail. Non-doped, n-doped, and p-doped Ge films with different thickness of 50, 100, 200, and 400 nm were studied. The capacity of non-doped sample decreased rapidly at high current densities compared to doped films. Also the conductivity of non-doped samples (>1.25 mS/cm) was too low. The n-doped Ge films showed best conductivity of 30 mS/cm, whereas p-doped only had 5 mS/cm. This increased the life cycle of n-doped films compared to p-doped and non-doped. They prepared electrodes with different thickness of 50, 100, 200, and 400 nm. The n-doped Ge film of thickness 200 nm exhibited best life cycle among others. It showed a stable discharge capacity of 780 $\mu Ahcm^2/cm$ over 180 cycles.

In this work, they opened up the future scope for Ge as an eminent anode for thin-film battery applications.

Liu et al. [25] studied and reported the electrochemical performance of macroporous germanium thin films developed by electrodeposition from ionic liquid. They did a comparative study between dense Ge and 3D-ordered macroporous (3DOM) films. 3DOM electrodes exhibited an improved rate performance and better life cycle compared to the other. At the end of 50 cycles, 3DOM film delivered a capacity of 844 mAh/g at 0.2C, whereas dense Ge film was only able to have a lithium storage of 611 mAh/g. Cracks formed on electrode surface of dense film led to pulverization of electrode and ended up in poor performance. Due to the porous structure of 3DOM, there was no crack formation. It played a role for minimizing the volume expansion and also increased the Li-ion diffusion pathways.

Abel et al. [27] reported the improved electrochemical performance of nanocolumnar $Ge_{0.9}Se_{0.1}$ electrodes prepared via vacuum deposition. They did an investigation on Se-free and Se-alloyed Ge as anode. The addition of Se to nanocolumnar-structured Ge increased its Li-ion transport and thereby high rate performance. A film with a thickness of 100 nm got completely delithiated in a very short time of 5 s. Cells were highly stable over 1000 cycles at a rate of 61 A/g with a retention of 70%. At 1342 A/g, it exhibited a capacity of 940 mAh/g with 75% retention. It is clear from the TEM images that during the first cycle, there formed a Li-Se high ionic conducting phase in the material that paved way for the enhanced performance.

3. Structural, morphological, and electrochemical results of thermally evaporated germanium thin films

3.1 Experimental

The deposition of thin-films was carried out in a cluster tool deposition system installed in the lab (**Figure 1** shows the photograph of the system). The cluster tool system is equipped with multiprocess chambers (including DC, RF-magnetron sputtering, and thermal evaporation) connected through load-lock and transfer chamber separated by gate valves (for independent operation). The system is equipped to transfer samples from one process chamber to the other using magnetic transfer arms without atmospheric exposure of the samples. All the process chambers are equipped with turbo-molecular pump and a roughing

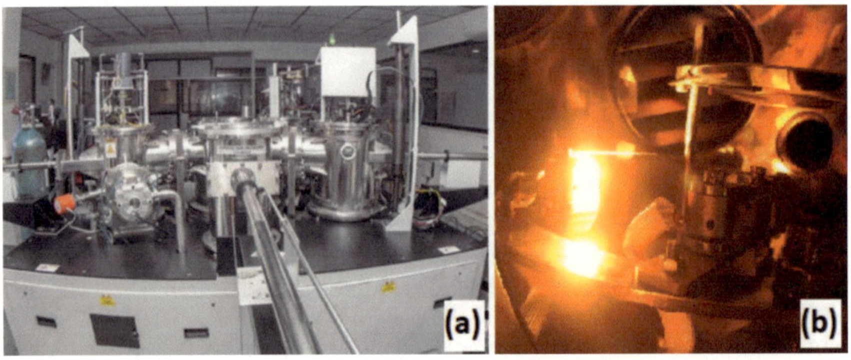

Figure 1.
(a) Physical vapor deposition system installed in the lab and (b) thermal evaporation chamber view during operation.

pump. For the present work, germanium thin films were deposited at room temperature on copper foil via thermal evaporation process with Ge granules as a source thermally evaporated from moly boat. Thermal evaporation chamber is loaded with a 5 × 5 cm^2 Cu foil (substrate) and pumped to a base vacuum of 5 × 10^{-7} mbar prior to the evaporation process. The thermal evaporator uses a power supply of 200 A and 10 V maximum. Once the base vacuum is attained, the current was gradually increased from 0 to 70 A, and the germanium granules were allowed to completely evaporate at 70 A by maintaining the same for a few minutes. During deposition the substrate holder was rotated at 10 rpm to ensure uniformity over the area of deposition. The deposited films underwent structural and morphological characterization as well as electrochemical analysis. X-ray diffraction of the resultant films was obtained from Cu Kα X-rays (Bruker AXS D8 Advance, Germany). Cross-section image was recorded using JSM 6490 (JEOL, Japan) scanning electron microscopy (SEM), while the surface images were recorded using JSM 7610FPlus (JEOL, Japan) field emission SEM microscope. Raman spectrum was collected using a 633 nm diode laser (Witech Alpha 300R, Germany). Surface analysis of the films was carried out with Axis Ultra spectrometer (Kartos, UK) with monochromated Al-K_α as X-ray source. The spectra obtained were calibrated based on C 1S peak at 284.6 eV. Galvanostatic charge-discharge performances were studied at room temperature using electrochemical workstation (Biologic Instruments, USA) with a voltage range of 0.01–2 V. Ge films deposited on Cu foil were cut into small pieces with a loading of 63 μg/cm^2 for electrochemical testing in coin cells (CR 2032) crimped inside the argon-filled glovebox maintained with O_2 and H_2O levels <1.0 ppm (UNIlab Pro, MBRAUN, Germany). Lithium foil was used as counter electrode with glass microfibers (GF/D, Whatman) as separators and 1 M $LiPF_6$ dissolved in EC: DMC (1:1 vol. %) as electrolyte (PuriEL, Canada). VC additive electrolytes were prepared in glovebox by adding 5 and 10 wt.% of VC to the electrolyte under stirring for a few minutes. Cycled electrodes (100 cycles) were washed in dimethyl carbonate (DMC) and dried in the glovebox before the surface chemical analysis.

3.2 Structural and morphological analysis

Figure 2a shows the XRD pattern of the deposited Ge films. It is shown that the films obtained are crystalline in nature. Characteristic peaks of Ge were located at 26.97°, 45.2°, and 54.8° that correspond to 111, 220, and 311, respectively [13]; small

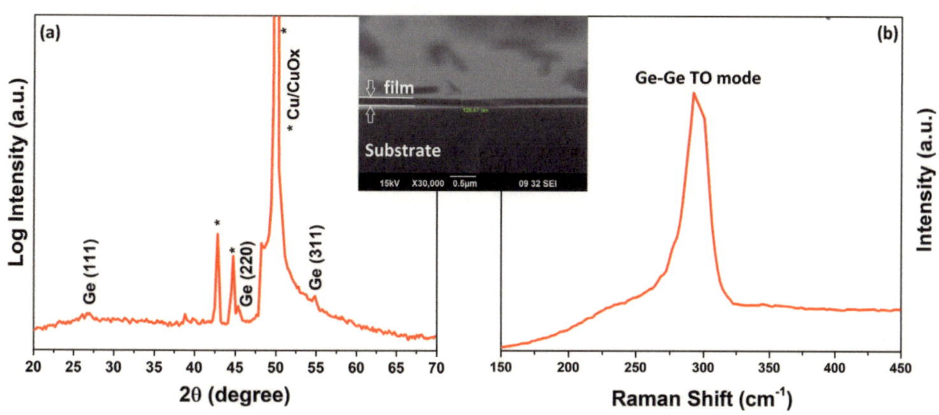

Figure 2.
(a) XRD of Ge film deposited on Cu-foil and (b) Raman spectrum of prepared Ge film. Inset shows the cross sectional SEM image of the Ge film.

shift in the peak positions could be due to strain effects. The presence of Cu/CuO_x (substrate) peaks is indexed in the figure [29]. Raman spectrum of Ge films with a peak at 292 cm^{-1} is depicted in **Figure 2b**, which represents the Ge-Ge transverse optical phonon mode. The sharp and intense peak confirms the local structural ordering of the film [22], while a small shift to lower frequency may be due to the strain effect. Inset in **Figure 2** displays the cross-sectional SEM image of Ge film which clearly shows Ge layer thickness of about 125 nm. It also reveals the high-thickness uniformity of the deposited thin film.

3.3 Electrochemical results

The additive concentration was varied, and sample identifications are based on the weight percentage of the additive, 0 wt.% VC (without additive), 5 wt.% VC and 10 wt.% VC Ge electrodes, respectively. For the understanding of electrochemical redox potentials of the deposited films, cyclic voltammetry was done at a scan rate of 0.1 mV/s in a voltage window of 0–1.2 V and is shown in **Figure 3**. **Figure 3a–c** corresponds to the 0 wt.% VC electrode, 5 wt.% VC, and 10 wt.% VC Ge electrodes, respectively. It is seen that for all three samples, there is a strong oxidation peak at 0.53 V due to the delithiation process. Reduction peaks were observed at 0.05, 0.14, 0.31, and 0.38 V that attributes to Li-Ge alloy formation [3, 15]. In second and third cycle, two reduction peaks got shifted to 0.38 and 0.50 V for all three electrodes. Structural modifications happened during the first lithiation which contributes to the peak shift in further cycles. Oxidation and reduction peaks are well superimposed for all, which reveals the reversibility of the films.

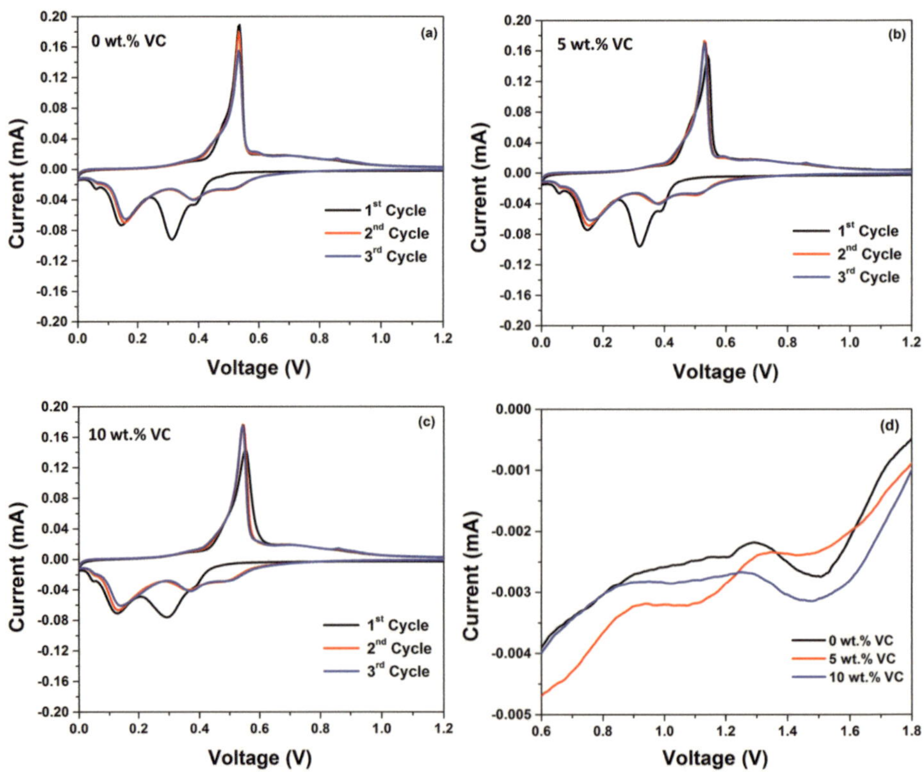

Figure 3.
Cyclic voltammogram of Ge films (a) without electrolyte additive (b) with 5 wt.% VC (c) with 10 wt.% VC and (d) first cycle reduction peaks of 0 wt.%, 5 wt.% and 10 wt.% VC samples in a voltage range of 0.6–1.8 V.

The first cycle reduction peaks of all the three samples in a voltage range of 0.6–1.8 V were shown in **Figure 3d**. The presence of peaks at 1.05 and 1.02 V in 5 wt. % and 10 wt.% VC films, respectively, is attributed to the reduction of VC [3]. 0 wt. %, 5 wt. %, and 10 wt.% VC films showed a reduction peak at 1.5 V. It may be due to the presence of SEI layer formed from surface oxides. Reduction of VC in 5 wt.% VC Ge film formed a stable layer on electrode/electrolyte interface that prevented it from further electrolyte decomposition which resulted in improved electrochemical performance.

Ge thin films were cycled at a current density of 1.6 A/g in a potential window of 0.01–2 V. **Figure 4a–c** shows the charge-discharge profiles of the germanium electrodes without and with electrolyte additives as indicated. **Figure 4d** depicts the life cycle of the material with and without using the electrolyte additive, VC. Ge electrodes with 5 wt.% VC and 10 wt.% VC show stable performance compared to the 0 wt.% VC. Unstable SEI layer formed on the electrode surface due to the electrolyte decomposition leads to capacity fade after a few cycles. It is seen that the 0 wt.% VC electrode delivered a capacity of 871 mAh/g after the first discharge and gradually capacity decreased to 386 mAh/g with only 50% retention at the end of 100 cycles. Cell with 5 wt.% VC as electrolyte additive showed improved capacity compared to that of 10 wt.% VC. A high initial capacity of 975 mAh/g with a retention of 94 and 99.4% Coulombic efficiency (CE) at the end of 100 cycles was recorded for the 5 wt.% electrode. Whereas for 10 wt.% VC as electrolyte additive, the electrode exhibited lower initial capacity of 778 mAh/g and a CE of 99.4% at the end of 100 cycles. At higher potential, VC got reduced and formed a stable layer on electrode/electrolyte interface that contains poly VC compounds. The layer formed has proper Li-ion conductivity and also prevents the electrolyte from decomposition. Improved performance of Ge electrode with 5 wt.% VC compared to 10 wt.% VC is attributed to the less amount of poly species formed on the electrode surface due to VC reduction.

Electrochemical impedance studies were done on 0 wt. % VC, 5 wt. % VC, and 10 wt.% VC electrodes for understanding the effect of electrolyte additive and is

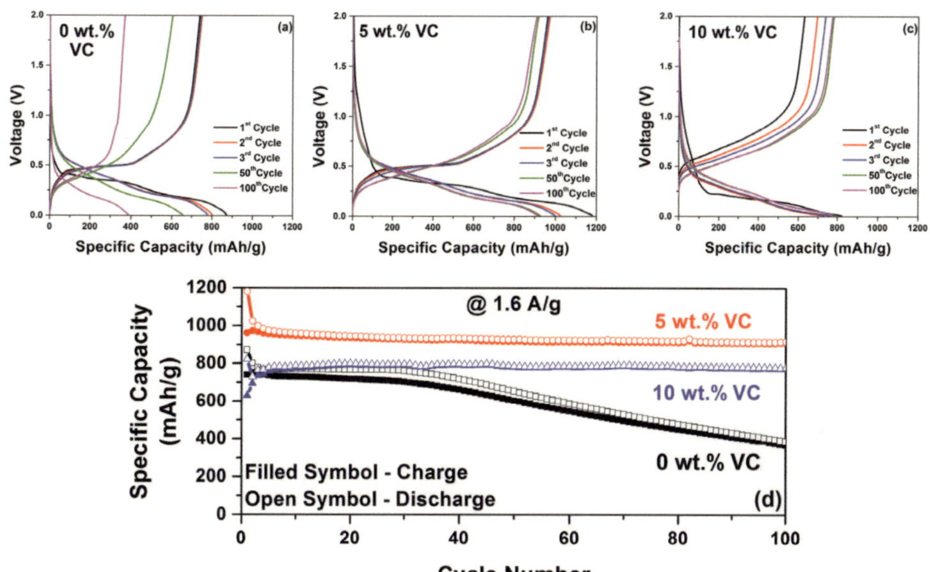

Figure 4.
Charge-discharge profiles of (a) Ge with 0 wt.% VC (b) Ge with 5 wt.% VC and (c) Ge with 10 wt.% VC (d) Cycling data of Ge electrodes with and without electrolyte additive.

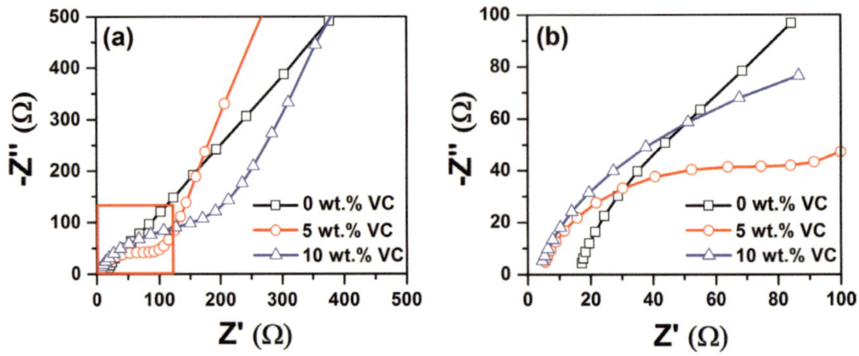

Figure 5.
(a) Electrochemical impedance spectroscopy of Ge films at OCV condition and (b) enlarged view of the region marked in 5(a).

illustrated in **Figure 5**. **Figure 5a** indicates the electrochemical impedance spectroscopy (EIS) results of the three electrodes. It is observed that except the 0 wt.% VC electrode, the electrodes with electrolyte additive show a depressed semicircle in the high-mid-frequency region followed by a straight line in the low-frequency region. 0 wt.% VC electrode shows only a straight line without any obvious semicircle indicating a capacitive charge storage with poor electronic conduction. The semicircle intersection with the x-axis at high frequency gives information about the series resistance, and information about charge-transfer resistance is given by the intersection at mid-frequency region. The slope of the straight line in the low-frequency region gives information about Li-diffusion kinetics. **Figure 5b** shows the closer view of high-frequency region marked in (a). Series resistance (R_S) value is high for 0 wt.% VC compared to 5 wt. % VC and 10 wt.% VC samples. It exhibits a R_S value of 17.15, 5.46, and 4.54 Ω for 0 wt.% VC, 5 wt.%, and 10 wt.% VC samples, respectively. Charge-transfer resistance, R_{ct} values of VC 5 wt.%, ad VC 10 wt.% electrodes were 106.0 and 237.8 Ω, respectively, while the 0 wt.% VC electrode did not show any semicircle indicating infinite R_{ct} value. Lower R_s ad R_{ct} values for the 5 wt.% VC additive electrode validate the improved electrochemical performance discussed above.

4. Surface chemical analysis of cycled germanium thin-film electrodes

In order to gain a clear insight on the solid-electrolyte interphase layer, surface chemical analysis was performed. XPS analysis were carried out on as-deposited as well as 0 wt.%, 5 wt.%, and 10 wt.% VC additive electrochemically cycled electrodes (in charged condition, after 100 cycles). **Figure 6a** shows the high-resolution spectra of C 1 s in which C-C peak at 284.65 eV which is common in all the samples. The numbers on the left side of the individual spectrum indicate the intensity multiplication factor for normalization. The presence of CO_3, C═O, and C-O were observed at 289.85, 288.05, and 286.75 eV, respectively, only in cycled electrodes due to electrolyte decomposition. Peak intensity of carbonates is less in 5 wt.% electrode compared with 0 wt. % VC and 10 wt.% VC sample. C-Ge peak at 282.75 eV is seen only in 0 wt.% VC sample, which implies the SEI layer formation during cycling owing to the unwanted side reactions. **Figure 6b** shows the O 1S spectra of the films having OH^- peak at 532.25 eV only for the as-deposited sample, whereas GeO_x is present in all the four samples at 531.45 eV. The intensity of GeO_x peak is high in cycled electrodes compared with the as-deposited. Li 1 s spectra are shown in

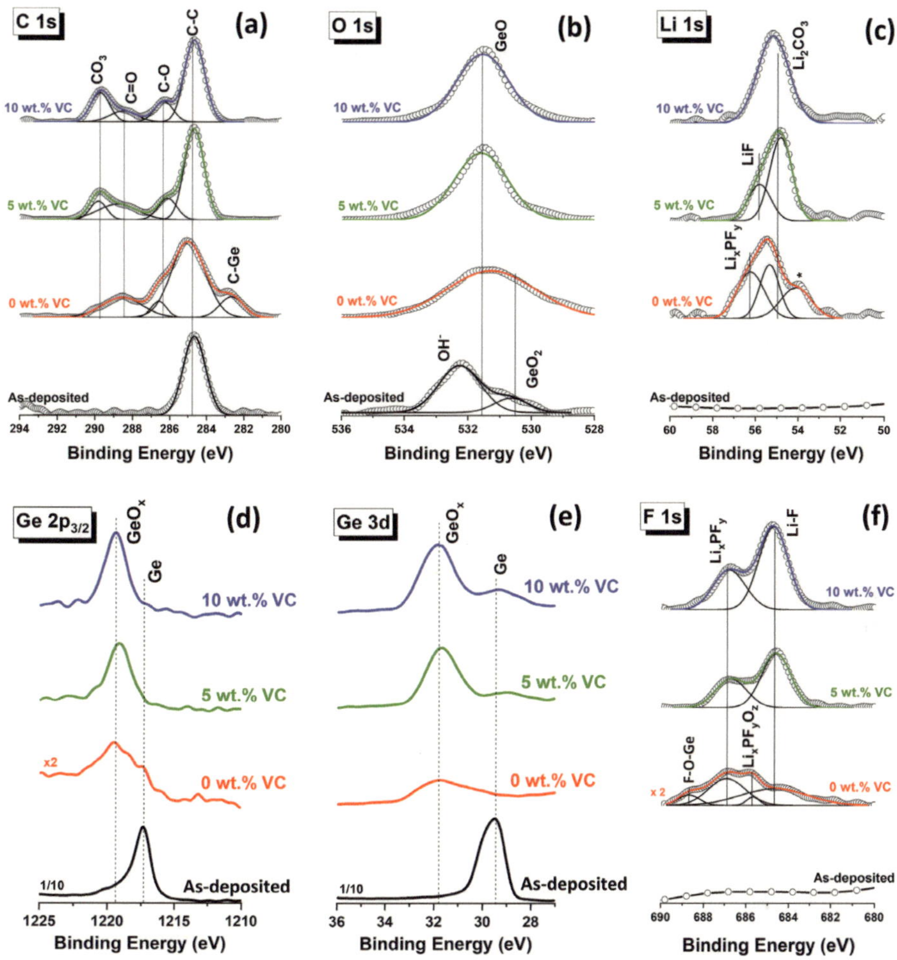

Figure 6.
High resolution XPS spectra of (a) C 1s, (b) O 1s, (c) Li 1s, (d) Ge 2p, (e) Ge 3d and (f) F 1s of as-deposited Ge film and cycled electrodes with 0 wt.%, 5 wt.% and 10 wt.% VC.

Figure 6c indicating the presence of LiF and Li_2CO_3 at 55.95 and 54.95 eV, respectively, except for the as-deposited sample. This may be from the SEI layer formed on the surface of the electrodes due to the decomposition of electrolyte at low voltages. It is seen that the amount of Li_2CO_3 is high in 10 wt.% VC electrode compared with 0 wt. % VC and 5 wt.% VC electrode. Peaks at 1219.35 and 1217.35 eV shown in **Figure 6d** correspond to GeO_x and Ge, respectively. In as-deposited sample, Ge exists only in its elemental form. However, Ge is mostly observed to be as GeO_x in cycled electrodes. It is indeed important to note that the Ge peak shoulder was for the 0 wt.% VC electrode in Ge $2p_{3/2}$. It is believed to be due to the effect of noise and signal fluctuation as the intensity for the whole spectrum was too low (the intensity reported for this electrode is multiplied by a factor of 2 as indicated in the figure). In general, Ge 3d is preferable over Ge 2p if Ge concentration is low or if it is expected to be a buried layer. Ge 3d electrons are capable of penetrating from buried layer compared to that of Ge 2p electrons due to its higher kinetic energy. This can be further verified by recording the high-resolution spectra in Ge 3d region. **Figure 6e** shows the high-resolution Ge 3d spectra with Ge and GeO_x peaks observed at 31.75 and 29.45 eV, respectively. The presence of Ge peak for the electrodes cycled with VC electrolyte additive indicates a thin SEI layer. In 0 wt.% VC electrode, only

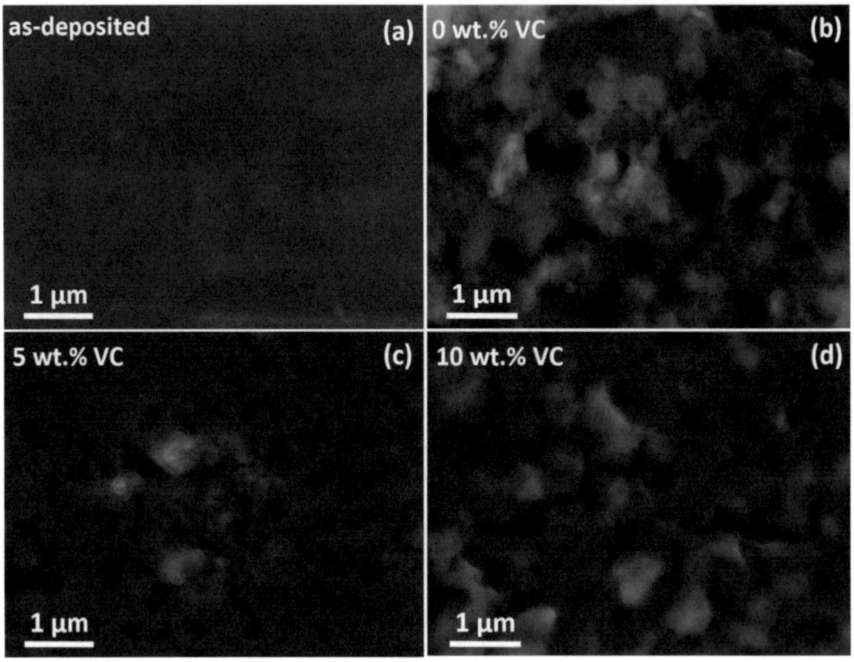

Figure 7.
FESEM images of the electrodes surface for (a) as-deposited Ge film, cycled electrodes with (b) 0 wt.%, (c) 5 wt.% and (d) 10 wt.% VC.

the presence of GeO_x is observed, which can be attributed to the thicker SEI layer formed by electrolyte decomposition during the cycling of electrodes [19]. Also Ge 3d seems to be more sensitive to elemental germanium than Ge 2p as observed from its intensity. The 5 wt.% and 10 wt.% VC electrodes have both Ge and GeO_x peaks that indicates thinner and stable SEI layer. The intensity of oxide peak is high in 10 wt.% VC electrode [30, 31]. F 1 s spectra is given in **Figure 6f** which depicts the presence of LiF, $Li_xPF_yO_z$, and Li_xPF_y at 684.8, 685.7, and 686.9 eV, respectively [32]. As expected, these peaks are absent in as-deposited sample. The presence of LiF and Li_xPF_y is seen more in 10 wt.% VC electrode, while a small amount of F-O-Ge (~688.6 eV) and $Li_xPF_yO_z$ are observed in 0 wt.% electrode only. Increased amount of LiF, Li_2CO_3, and GeO_2 in 10 wt.% VC electrode can act as an insulating layer for Li-ion diffusion that attenuates the electrochemical performance compared with 5 wt.% VC electrode. Further to investigate the surface morphology of the as-deposited and cycled electrodes (after 100 cycles), SEM images were recorded; **Figure 7** shows the SEM images of as-deposited, 0 wt. % VC, 5 wt. % VC, and 10 wt.% VC electrodes. As-deposited films have smooth, pore-free, and crack-free surface. It is observed that carbon-rich SEI layer along with pores and cracks have been formed on the surface of 0 wt.% VC electrode compared to the other two electrodes, while 10 wt.% VC electrode showed more porous and cracked surface compared to 5 wt.% VC electrode. This along with the XPS analysis indicates that 5 wt.% VC electrode high-performance anode is well correlated with the thin and stable SEI layer.

5. Conclusions

Ge thin-film anodes were developed via thermal evaporation process for LIB applications. Structural analysis revealed the crystalline nature of the deposited Ge film. The effect of VC as electrolyte additive and its concentration on electrochemical cycling stability were studied in detail. Galvanostatic charge-discharge studies showed best performance for 5 wt.% VC compared to that of 0 and 10 wt.% VC. It exhibited a high initial capacity of 975 mAh/g with a retention of 94 and 99% Coulombic efficiency at the end of 100 cycles. Cyclic voltammetry clearly showed the reduction of VC at 1.05 V in 5 wt.% additive sample that led to the formation of a stable and thin SEI layer on the film surface which prevented further electrolyte decomposition. This is attributed to the low-series and charge-transfer resistance values for 5 wt.% VC additive. Improved electrochemical performance of the electrode cycled with 5 wt.% VC was correlated with the surface chemical analysis of solid-electrolyte interphase layer. XPS results confirmed that the SEI layer had a mixed organic and inorganic composition for all the electrodes. The SEI layer composition with VC as electrolyte additive had LiF as a major component and significant amount of Li_xPF_y; in contrary, it was $Li_xPF_yO_z$ and negligible LiF for the electrode without VC additive. The presence of Li_2CO_3 in all the electrodes indicated that it played a least significant role in the performance. The surface SEM images of the cycled electrodes were consistent with the electrochemical performance and its correlation with the XPS results.

Acknowledgements

DS acknowledges Science and Engineering Research Board (SERB), India, for the award of Ramanujan Fellowship for supporting this work (SB/S2/RJN-100/2014) and financial support to SSJ. The infrastructural support and the PVD cluster deposition system funded by Amrita Vishwa Vidyapeetham are gratefully acknowledged.

Conflict of interest

The authors declare no conflict of interest.

Author details

Silpasree S. Jayasree, Shantikumar Nair and Dhamodaran Santhanagopalan*
Centre for Nanosciences and Molecular Medicine, Amrita Vishwa Vidyapeetham, AIMS, Kochi, Kerala, India

*Address all correspondence to: dsgopalan20710@aims.amrita.edu

References

[1] Bogart TD, Chockla AM, Korgel BA. High capacity lithium ion battery anodes of silicon and germanium. Current Opinion in Chemical Engineering. 2013;**2**:286-293. DOI: 10.1016/j.coche.2013.07.001

[2] Chan CK, Zhang XF, Cui Y. High capacity Li ion battery anodes using Ge nanowires. Nano Letters. 2008;**8**:307-309. DOI: 10.1021/nl0727157

[3] Gulzar U, Li T, Bai X, Goriparti S. Nitrogen-doped single walled carbon nanohorns enabling effective utilization of Ge nanocrystals for next generation lithium ion batteries. Electrochimica Acta. 2018;**298**:89-96. DOI: 10.1016/j.electacta.2018.11.130

[4] Tian H, Xin F, Wang X. High capacity group-IV elements (Si, Ge, Sn) based anodes for lithium-ion batteries. Journal of Materiomics. 2015;**1**:153-169. DOI: 10.1016/j.jmat.2015.06.002

[5] Chen L, Wang K, Xie X. Effect of vinylene carbonate (VC) as electrolyte additive on electrochemical performance of Si film anode for lithium ion batteries. Journal of Power Sources. 2007;**174**:538-543. DOI: 10.1016/j.jpowsour.2007.06.149

[6] Graetz J, Ahn CC, Yazami R. Nanocrystalline and thin film germanium electrodes with high lithium capacity and high rate capabilities. Journal of the Electrochemical Society. 2004;**151**:A698-A702. DOI: 10.1149/1.1697412

[7] Ocon JD, Lee KJ, Lee J. High energy density germanium anodes for next generation lithium ion batteries. Applied Chemical Engineering. 2014;**251**:1-13. DOI: 10.14478/ace.2014.1008

[8] DiLeo RA, Ganter MJ, Landi BJ. Germanium–single-wall carbon nanotube anodes for lithium ion batteries. Journal of Materials Research. 2010;**25**:1441-1446. DOI: 10.1557/JMR.2010.0184

[9] Yoon S, Park CM, Sohn HJ. Electrochemical characterizations of germanium and carbon-coated germanium composite anode for lithium-ion batteries. Electrochemical and Solid-State Letters. 2008;**11**:A42-A45. DOI: 10.1149/1.2836481

[10] Ren JG, Wu QH, Tang H. Germanium–graphene composite anode for high-energy lithium batteries with long cycle life. Journal of Materials Chemistry A. 2013;**1**:1821-1826. DOI: 10.1039/c2ta01286c

[11] Hu Z, Zhang S, Zhang C. High performance germanium-based anode materials. Coordination Chemistry Reviews. 2016;**326**:34-85. DOI: 10.1016/j.ccr.2016.08.002

[12] Yu Y, Yan C, Gu L. Three-dimensional (3D) bicontinuous Au/amorphous-Ge thin films as fast and high-capacity anodes for lithium-ion batteries. Advanced Energy Materials. 2013;**3**:281-285. DOI: 10.1002/aenm.201200496

[13] Liu X, Lin N, Cai W. Mesoporous germanium nanoparticles synthesized in molten zinc chloride at low temperature as a high-performance anode for lithium-ion batteries. Dalton Transactions. 2018;**47**:7402-7406. DOI: 10.1039/C8DT01060A

[14] Park MH, Cho Y, Kim K. Germanium nanotubes prepared by using the Kirkendall effect as anodes for high-rate lithium batteries. Angewandte Chemie, International Edition. 2011;**50**:9647-9650. DOI: 10.1002/anie.201103062

[15] Wang X, Susantyoko RA, Fan Y. Vertically aligned CNT-supported thick

Ge films as high-performance 3D anodes for lithium ion batteries. Small. 2014;**10**:2826-2829. DOI: 10.1002/smll.201400003

[16] Rudawski NG, Darby BL, Yates BR. Nanostructured ion beam-modified Ge films for high capacity Li ion battery anodes. Applied Physics Letters. 2012;**100**:083111. DOI: 10.1063/1.3689781

[17] Baggetto L, Notten PH. Lithium-ion (de) insertion reaction of germanium thin-film electrodes: An electrochemical and in situ XRD study. Journal of the Electrochemical Society. 2009;**156**:A169-A175. DOI: 10.1149/1.3055984

[18] Haregewoin AM, Wotango AS, Hwang BJ. Electrolyte additives for lithium ion battery electrodes: Progress and perspectives. Energy and Environmental Science. 2016;**9**:1955-1988. DOI: 10.1039/c6ee00123h

[19] Mishra K, Xu W, Engelhard MH. The effect of solvent on the capacity retention in a germanium anode for lithium ion batteries. Journal of Electrochemical Energy Conversion and Storage. 2018;**15**:041012. DOI: 10.1115/1.4039860

[20] Li X, Yang Z, Fu Y. Germanium anode with excellent lithium storage performance in a germanium/lithium–cobalt oxide lithium-ion battery. ACS Nano. 2015;**9**:1858-1867. DOI: 10.1021/nn506760p

[21] Laforge B, Levan-Jodin L, Salot R. Study of germanium as electrode in thin-film battery. Journal of the Electrochemical Society. 2008;**155**:A181-A188. DOI: 10.1149/1.2820666

[22] Hwang CM, Park JW. Electrochemical characterizations of multi-layer and composite silicon–germanium anodes for Li-ion batteries using magnetron sputtering. Journal of Power Sources. 2011;**196**:6772-6780. DOI: 10.1016/j.jpowsour.2010.10.061

[23] Susantyoko RA, Wang X, Sun L. Influences of annealing on lithium-ion storage performance of thick germanium film anodes. Nano Energy. 2015;**12**:521-527. DOI: 10.1016/j.nanoen.2015.01.024

[24] Al-Obeidi A, Kramer D, Mönig R. Mechanical stresses and crystallization of lithium phosphorous oxynitride-coated germanium electrodes during lithiation and delithiation. Journal of Power Sources. 2016;**306**:817-825. DOI: 10.1016/j.jpowsour.2015.12.057

[25] Liu X, Zhao J, Hao J. 3D ordered macroporous germanium fabricated by electrodeposition from an ionic liquid and its lithium storage properties. Journal of Materials Chemistry A. 2013;**1**:15076. DOI: 10.1039/c3ta12923c

[26] Abel PR, Chockla AM, Lin YM. Nanostructured $Si_{(1-x)}Ge_x$ for tunable thin film lithium-ion battery anodes. ACS Nano. 2013;**7**:2249-2257. DOI: 10.1021/nn3053632

[27] Abel PR, Klavetter KC, Heller A. Thin nanocolumnar $Ge_{0.9}Se_{0.1}$ films are rapidly lithiated/delithiated. The Journal of Physical Chemistry C. 2014;**118**:17407-17412. DOI: 10.1021/jp504327e

[28] Abel PR, Klavetter KC, Jarvis K. Sub-stoichiometric germanium sulfide thin-films as a high-rate lithium storage material. Journal of Materials Chemistry A. 2014;**2**:19011-19018. DOI: 10.1039/C4TA04496G

[29] Cheng SL, Chen MF. Fabrication, characterization, and kinetic study of vertical single-crystalline CuO nanowires on Si substrates. Nanoscale Research Letters. 2012;**7**:119. DOI: 10.1186/1556-276X-7-119

[30] Gangaja B, Chandrasekharan S, Vadukumpully S. Surface chemical analysis of CuO nanofiber composite electrodes at different stages of lithiation/delithiation. Journal of Power Sources. 2017;**340**:356-364. DOI: 10.1016/j.jpowsour.2016.11.087

[31] Pumera M, Iwai H, Miyahara Y. Germanium-oxide-coated carbon nanotubes. Nanotechnology. 2009;**20**:425606. DOI: 10.1088/0957-4484/20/42/425606

[32] John J, Gangaja B, Nair SV. Conformal coating of TiO2 shell on silicon nanoparticles for improved electrochemical performance in Li-ion battery applications. Electrochimica Acta. 2017;**235**:191-199. DOI: 10.1016/j.electacta.2017.03.127